Toxic Pollutants in Water: Health Risk Assessment and Removal

Toxic Pollutants in Water: Health Risk Assessment and Removal

Guest Editors

Shakeel Ahmad
Shicheng Zhang
Mujtaba Baqar
Eric Danso-Boateng

MDPI

Basel • Beijing • Wuhan • Barcelona • Belgrade • Novi Sad • Cluj • Manchester

Guest Editors
Shakeel Ahmad
Kunming University of Science and Technology
Kunming
China

Shicheng Zhang
Fudan University
Shanghai
China

Mujtaba Baqar
Government College University Lahore
Lahore
Pakistan

Eric Danso-Boateng
University of Leeds
Leeds
UK

Editorial Office
MDPI AG
Grosspeteranlage 5
4052 Basel, Switzerland

This is a reprint of the Special Issue, published open access by the journal *Water* (ISSN 2073-4441), freely accessible at: https://www.mdpi.com/journal/water/special_issues/9VO26H2ZR3.

For citation purposes, cite each article independently as indicated on the article page online and as indicated below:

Lastname, A.A.; Lastname, B.B. Article Title. *Journal Name* **Year**, *Volume Number*, Page Range.

ISBN 978-3-7258-5687-9 (Hbk)
ISBN 978-3-7258-5688-6 (PDF)
https://doi.org/10.3390/books978-3-7258-5688-6

Contents

About the Editors

Shakeel Ahmad

Shakeel Ahmad obtained his Ph.D. in environmental engineering from Fudan University, China, in 2020. In 2020–2023, he worked as a postdoctoral researcher at Nankai University, China. Currently, he is working as a professor at the Faculty of Environmental Science and Engineering, Kunming University of Science and Technology, China. His research interests are water treatment and soil remediation, specifically, the removal of heavy metals and organic pollutants through adsorption, catalytic reduction, and advanced oxidation processes. Prof. Shakeel has published three book chapters and over 40 journal articles with over 1700 citations (h-index 24). Moreover, he has served as a Guest Editor of MDPI *Toxics* (2024-25) and MDPI *Water* (2023–24).

Shicheng Zhang

Shicheng Zhang is a full professor at the Department of Environmental Science and Engineering, Fudan University, Shanghai, China, and the director of the Shanghai Technical Service Platform for Pollution Control and Resource Utilization of Organic Wastes. He currently serves as a core group member of the Sustainable Waste Management Program for the Association of Pacific Rim Universities (APRU SWMP), a scientific committee member of the World Society of Engineering Thermochemistry (WSETC), and a fellow member of the Australasia Practical Zero Emissions Society (APZES). His research group focuses on biomass waste and organic waste utilization by chemical, biological and biochemical methods, especially the hydrothermal conversion of organic waste to high-value-added products. Prof. Zhang has published 9 book chapters and more than 250 journal articles, with SCI citations over 15000 times (h-index 69). He has received several awards for his scholarly achievements, such as the Second Prize of Natural Science Award for Scientific Research in Colleges and Universities of the Ministry of Education, the Shanghai Oriental Talents Program—Top-notch Talents Sub-Program, and so on. He serves as an Associate Editor of *Environmental Engineering Research* and *NPJ Materials Sustainability*. He has been recognized as one of the World's Top 2% Scientists and a Highly Cited Researcher.

Mujtaba Baqar

Mujtaba Baqar is a committed scholar and accomplished researcher in the field of Environmental Sciences, with over 13 years of academic and research experience. Since 2018, he has been serving as an Assistant Professor at the Sustainable Development Study Centre (SDSC), Government College University, Lahore, Pakistan. A recipient of three academic gold medals, Dr. Mujtaba has authored 56 journal articles and four book chapters, achieving h-index of 21. He completed his postdoctoral fellowship at Nankai University, China, where he advanced his expertise in environmental pollutant analysis and risk assessment. He has successfully secured competitive research grants totaling PKR 16 million. In recognition of his teaching excellence, he was honored with the Best University Teacher Award by the Higher Education Commission (HEC) of Pakistan. He has served as the youngest elected member of the Senate at the University of the Punjab and as an elected member of the Academic Council at Government College University, Lahore.

Eric Danso-Boateng

Eric Danso-Boateng holds a PhD in Chemical Engineering from Loughborough University, UK. He is presently a Teaching Fellow at the School of Process and Chemical Engineering, University of Leeds, UK. He previously worked as a University Teacher in the Department at Loughborough University, UK, and a Senior Lecturer at Kumasi Technical University, Ghana. His research interest focuses on waste-to-energy; biofuels; resource recovery from waste; biomass conversion technologies; carbon nanomaterials from waste and applications in pollutant control; and water quality and health risk assessment. Dr. Eric has published a patent, two books, three book chapters, and over 20 journal articles with over 1,285 citations (h-index 17). Moreover, he is an Associate Editor of *Sustainable & Green Materials and Pollution Study*, and has served as a Guest Editor of MDPI *Sustainability* (2023) and MDPI *Water* (2023).

water

MDPI

Editorial

Toxic Pollutants in Water: Health Risk Assessment and Removal

Shakeel Ahmad [1,2,*], Shicheng Zhang [3,4], Mujtaba Baqar [5] and Eric Danso-Boateng [6]

1 Yunnan Provincial Key Lab of Soil Carbon Sequestration and Pollution Control, Faculty of Environmental Science and Engineering, Kunming University of Science and Technology, Kunming 650500, China
2 Yunnan International Joint Laboratory for Emission Reduction and Carbon Sequestration in Agricultural Soils, Kunming 650500, China
3 Shanghai Technical Service Platform for Pollution Control and Resource Utilization of Organic Wastes, Shanghai Key Laboratory of Atmospheric Particle Pollution and Prevention (LAP[3]), Department of Environmental Science and Engineering, Fudan University, Shanghai 200438, China; zhangsc@fudan.edu.cn
4 Shanghai Institute of Pollution Control and Ecological Security, Shanghai 200092, China
5 Sustainable Development Study Centre, Government College University, Lahore 54000, Pakistan; mujtababaqar@gcu.edu.pk
6 School of Chemical and Process Engineering, University of Leeds, Leeds LS2 9JT, UK; e.danso-boateng@leeds.ac.uk
* Correspondence: shakeel@kust.edu.cn; Tel.: +86-18621073050

1. Introduction

Clean water is a fundamental human right; however, it is increasingly under threat from toxic pollutants that infiltrate rivers, lakes, groundwater, and even treated drinking water supplies [1]. As industrialization accelerates and agricultural practices intensify, these contaminants originate from multiple sources, such as industrial effluents, agricultural runoff, improper waste disposal, aging infrastructure, and household products [2,3]. Once released into environment, many of these pollutants do not degrade easily and can bioaccumulate in aquatic organisms, eventually entering the human food chain and causing global public health crises [4,5]. Challenges lie not only in identifying these toxins but also in assessing their risks to human health and implementing effective strategies for their removal [6]. Water pollution is no longer confined to visible signs, such as oil slicks or floating debris. Modern pollutants are often invisible, insidious, and persistent in nature. Chronic occurrence of toxic pollutants, including heavy metals (HMs; Pb, Ni, As, Hg, Cd, Co, and Cu), dyes, and organic contaminants such as pharmaceutical residues, pesticides, and perfluoroalkyl and poly-fluoroalkyl substances (PFAS; perfluorooctanoic acid (PFOA) and perfluorosulfonic acid (PFOS)), in water is a significant environmental concern due to their persistence, toxicity, and potential risks to humans and ecosystems [7–9].

The health consequences of long-term exposure to low levels of toxic pollutants are profound [10]. Unlike acute illnesses caused by pathogens, the effects of chemical contamination often manifest over years or decades. Exposure to HMs has been linked to neurological disorders, kidney damage, developmental delays in children, and cancer [11]. Organic compounds and PFAS, known as forever chemicals, are associated with immune dysfunction, hormonal disruption, and increased cancer risk [12,13]. In many developing regions, contaminated water remains a primary source of disease and mortality, particularly among children [14]. Assessing the health risks posed by these pollutants requires robust scientific methodologies.

One of the major hurdles in managing water pollution is the sheer diversity and evolving nature of the contaminants. Regulatory frameworks often lag behind scientific discoveries, leaving many emerging pollutants unregulated. For instance, while some

 https://doi.org/10.3390/w17131896

countries have set limits on certain organic compounds, thousands of them remain unmonitored. Moreover, detection technologies must keep pace with this rapidly evolving landscape. Traditional water testing methods may not be sufficiently sensitive to detect trace amounts of emerging contaminants. Advanced analytical techniques can help to bridge this gap; however, their widespread implementation remains costly and technically demanding. Addressing the issues of toxic pollutants in water requires a multi-pronged approach that combines prevention, treatment, and policy reform [7]. Adsorption with biochar and activated carbon [15,16], reverse osmosis [17], membrane filtration [18], bioremediation [19], and advanced oxidation processes [20] have proven effective in removing a wide range of contaminants. Nanotechnology [21] and (bio)remediation [22], which use microbes or plants to absorb or break down toxins, are promising frontiers in sustainable water purification. Furthermore, community-based solutions, such as rainwater harvesting, decentralized filtration systems, and green infrastructure, can help to reduce reliance on centralized water treatment plants and mitigate contamination at the local level.

As an effort to promote sustainable water management and environmental protection, this Special Issue, titled "Toxic Pollutants in Water: Health Risk Assessment and Removal", of *Water* aimed to advance our understanding and remediation of water contamination by bringing together multidisciplinary research on pollutant sources, health risks, and innovative removal techniques. This Special Issue covered a range of topics related to toxic pollutant (HMs, dyes, and organic contaminants) contamination in water, approaches for toxic pollutant sources, environmental fate, health risk assessment, environmental composites/nanoparticles for toxic pollutant removal, and water treatment techniques, i.e., adsorption, desalination, catalytic reduction, advanced oxidation, bioremediation, and membrane filtration.

2. Findings Reported in This Special Issue

This Special Issue comprises nine articles, including eight research articles and one review article. The topic of "Toxic Pollutants in Water: Health Risk Assessment and Removal" is discussed in the following articles: Ikizoglu et al. (2024) investigated the pollution of two PFAS types, namely PFOA and PFOS, both of which are banned by the Stockholm Convention, in surface waters and surface water fish in the most densely populated and industrial region in Turkey [contribution 1]. Shaffique et al. (2024) explored research developments in migratory water birds, focusing on indicators of HM (Ni, Cu, Co, Zn, Pb, Cd, and Mn) pollution in the inland wetland resources of Punjab, Pakistan [contribution 2]. Jing et al. (2024) evaluated the ecological impact of wastewater discharges on microbial and contaminant dynamics in rivers to optimize wastewater treatment processes to better comply with Chinese environmental quality standards [contribution 3]. Piccardo et al. (2024) assessed the first evidence of the water bioremediation potential of *Ficopomatus enigmaticus* (Fauvel 1923), both living and dead, to remove contaminants and enhance water quality [contribution 4]. Medina Salas et al. (2023) explored the ZnO-CuO nanocomposite's potential as an efficient adsorbent for As(III) removal from water, showing that the straightforward and energy-efficient production of nanocomposite makes it promising for real-world water treatments [contribution 5]. Alhamzah et al. (2023) studied the control of bromate, a potentially carcinogenic disinfection by-product; its formation in desalinated seawater production; and its transmission with ammoniation in Makkah, Saudi Arabia [contribution 6]. Cáceda Quiroz et al. (2023) studied cyanide bioremediation carried out by *Bacillus subtilis* under alkaline conditions using cyanide concentrations and experimental conditions demonstrative of real mining wastewaters [contribution 7]. Mahringer et al. (2023) evaluated the redox behavior of Cr with redox-active substances (O_2, NO_3^-, Fe^{2+}, MnO_2) in reduction, coagulation, and biotic filtration drinking water treatment at pilot-scale [contribution 8]. Liang et al. (2023) reviewed the literature on the pollution components of pulp and paper wastewater, their environmental and health impacts, and the sustainable treatment, recycling, and utilization of pulp and paper wastewater [contribution 9].

3. Conclusions and Future Directions

Toxic pollutants in water represent a complex and urgent challenge that demands immediate attention from scientists, policymakers, industry leaders, and civil society. Researchers have developed eco-friendly and cost-effective materials and methods for the efficient removal of pollutants from contaminated water, wastewater regeneration and reuse, and reducing toxicant levels, paving the way for scalable and adaptable wastewater treatment and sustainable water management [23–25]. Although progress has been made, much work remains to be completed to ensure that everyone has access to safe and clean water. As we face the growing pressures of climate change, population growth, and resource scarcity, protecting our water resources is not just an environmental imperative; it is also a moral one. The health of future generations depends on the choices that we make today. Let us act decisively, collaboratively, and with a shared commitment to preserving our most vital natural resources.

Effective regulation is essential to curb the release of toxic pollutants at source. Governments must enforce stringent industrial discharge standards, invest in wastewater treatment infrastructure, and promote sustainable agricultural practices. Transparency in reporting water-quality data and setting enforceable maximum contaminant levels are critical steps toward ensuring accountability. International cooperation is equally important in this regard. Water pollution knows no borders, especially in transboundary river basins and coastal waters. Agreements such as the Stockholm Convention on Persistent Organic Pollutants [26] and the Paris Agreement [27] provide a framework for global action, but stronger enforcement and funding mechanisms are needed.

The future of water pollution control depends on our capacity to innovate, collaborate, and educate. By leveraging cutting-edge technologies (such as artificial intelligence and the Internet of Things), responding to challenges posed by climate change, modernizing governance frameworks, and increasing public awareness, we can build a robust and sustainable approach for protecting surface water and groundwater resources. As we progress, it is crucial to remain flexible and proactive in addressing new and evolving threats. Emphasizing integrated strategies that combine scientific research, community engagement, and informed policy decisions will be key to ensuring the long-term health and resilience of essential water systems and resources.

Author Contributions: Conceptualization, S.A., S.Z., E.D.-B. and M.B.; investigation, S.A.; writing—original draft preparation, S.A.; writing—review and editing, S.Z. and M.B.; project administration, S.Z.; data curation, S.A. and E.D.-B. All authors have read and agreed to the published version of the manuscript.

Funding: This research received no external funding.

Data Availability Statement: Not applicable.

Acknowledgments: We acknowledge all the editors, authors, and reviewers who contributed to this Special Issue.

Conflicts of Interest: The authors declare no conflicts of interest.

List of Contributions:

1. Ikizoglu, B. PFOA and PFOS Pollution in Surface Waters and Surface Water Fish. *Water* **2024**, *16*, 2342.
2. Shaffique, S.; Kang, S.-M.; Ashraf, M.A.; Umar, A.; Khan, M.S.; Wajid, M.; Al-Ghamdi, A.A.; Lee, I.-J. Research Progress on Migratory Water Birds: Indicators of Heavy Metal Pollution in Inland Wetland Resources of Punjab, Pakistan. *Water* **2024**, *16*, 1163.
3. Jing, W.; Sajnani, S.; Zhou, M.; Zhu, H.; Xu, Y. Evaluating the ecological impact of wastewater discharges on microbial and contaminant dynamics in rivers. *Water* **2024**, *16*, 377.
4. Piccardo, M.; Vellani, V.; Anselmi, S.; Bentivoglio, T.; Provenza, F.; Renzi, M.; Bevilacqua, S. The First Evidence of the Water Bioremediation Potential of *Ficopomatus enigmaticus* (Fauvel 1923): From Threat to Resource? *Water* **2024**, *16*, 368.
5. Medina Salas, J.P.; Gamarra Gómez, F.; Sacari Sacari, E.J.; Lanchipa Ramos, W.O.; Tamayo Calderón, R.M.; Mamani Flores, E.; Yapuchura Platero, V.; Florez Ponce de León, W.D.; Sandoval, E.M.L. ZnO-CuO nanocomposite as an efficient adsorbent for As (III) removal from water. *Water* **2023**, *15*, 4318.
6. Alhamzah, A.A.; Alofi, A.S.; Abid, A.A.; Fellows, C.M. Control of bromate formation in desalinated seawater production and transmission with ammoniation. *Water* **2023**, *15*, 3858.
7. Cáceda Quiroz, C.J.; Fora Quispe, G.d.L.; Carpio Mamani, M.; Maraza Choque, G.J.; Sacari Sacari, E.J. Cyanide bioremediation by *Bacillus subtilis* under alkaline conditions. *Water* **2023**, *15*, 3645.
8. Mahringer, D.; Zerelli, S.S.; Ruhl, A.S. Redox Behavior of Chromium in the Reduction, Coagulation, and Biotic Filtration (RCbF) Drinking Water Treatment—A Pilot Study. *Water* **2023**, *15*, 3363.
9. Liang, X.; Xu, Y.; Yin, L.; Wang, R.; Li, P.; Wang, J.; Liu, K. Sustainable utilization of pulp and paper wastewater. *Water* **2023**, *15*, 4135.

References

1. Mishra, R.K. Fresh water availability and its global challenge. *Br. J. Multidiscip. Adv. Stud.* **2023**, *4*, 1–78. [CrossRef]
2. Fida, M.; Li, P.; Wang, Y.; Alam, S.K.; Nsabimana, A. Water contamination and human health risks in Pakistan: A review. *Expo. Health* **2023**, *15*, 619–639. [CrossRef]
3. Singh, J.; Yadav, P.; Pal, A.K.; Mishra, V. Water pollutants: Origin and status. In *Sensors in Water Pollutants Monitoring: Role of Material*; Springer: Singapore, 2020; pp. 5–20.
4. Madhav, S.; Ahamad, A.; Singh, A.K.; Kushawaha, J.; Chauhan, J.S.; Sharma, S.; Singh, P. Water pollutants: Sources and impact on the environment and human health. In *Sensors in Water Pollutants Monitoring: Role of Material*; Springer: Singapore, 2020; pp. 43–62.
5. Fromme, H.; Küchler, T.; Otto, T.; Pilz, K.; Müller, J.; Wenzel, A. Occurrence of phthalates and bisphenol A and F in the environment. *Water Res.* **2002**, *36*, 1429–1438. [CrossRef] [PubMed]
6. Ahmad, S.; Liu, L.; Zhang, S.; Tang, J. Nitrogen-doped biochar (N-doped BC) and iron/nitrogen co-doped biochar (Fe/N co-doped BC) for removal of refractory organic pollutants. *J. Hazard. Mater.* **2023**, *446*, 130727. [CrossRef]
7. Rathi, B.S.; Kumar, P.S.; Vo, D.-V.N. Critical review on hazardous pollutants in water environment: Occurrence, monitoring, fate, removal technologies and risk assessment. *Sci. Total Environ.* **2021**, *797*, 149134. [CrossRef] [PubMed]
8. Okoro, H.K.; Orosun, M.M.; Oriade, F.A.; Momoh-Salami, T.M.; Ogunkunle, C.O.; Adeniyi, A.G.; Zvinowanda, C.; Ngila, J.C. Potentially toxic elements in pharmaceutical industrial effluents: A review on risk assessment, treatment, and management for human health. *Sustainability* **2023**, *15*, 6974. [CrossRef]
9. Wang, Z.; Luo, P.; Zha, X.; Xu, C.; Kang, S.; Zhou, M.; Nover, D.; Wang, Y. Overview assessment of risk evaluation and treatment technologies for heavy metal pollution of water and soil. *J. Clean. Prod.* **2022**, *379*, 134043. [CrossRef]
10. Manyepa, P.; Gani, K.M.; Seyam, M.; Banoo, I.; Genthe, B.; Kumari, S.; Bux, F. Removal and risk assessment of emerging contaminants and heavy metals in a wastewater reuse process producing drinkable water for human consumption. *Chemosphere* **2024**, *361*, 142396. [CrossRef]
11. Jafarzadeh, N.; Heidari, K.; Meshkinian, A.; Kamani, H.; Mohammadi, A.A.; Conti, G.O. Non-carcinogenic risk assessment of exposure to heavy metals in underground water resources in Saraven, Iran: Spatial distribution, monte-carlo simulation, sensitive analysis. *Environ. Res.* **2022**, *204*, 112002. [CrossRef]
12. Wang, L.; Xu, X.; Zhu, D.; Li, X.; Zhang, G.; Cheng, G.; Liu, Q.; Lin, L.; Wang, D. The occurrence and ecological risk assessment of typical organic polutants in the effluents of urban reclaimed water plants. *Acta Sci. Circumstantiae* **2021**, *41*, 1910–1919.

13. Ankley, G.T.; Cureton, P.; Hoke, R.A.; Houde, M.; Kumar, A.; Kurias, J.; Lanno, R.; McCarthy, C.; Newsted, J.; Salice, C.J. Assessing the ecological risks of per-and polyfluoroalkyl substances: Current state-of-the science and a proposed path forward. *Environ. Toxicol. Chem.* **2021**, *40*, 564–605. [CrossRef]
14. Kim, B.-N.; Cho, S.-C.; Kim, Y.; Shin, M.-S.; Yoo, H.-J.; Kim, J.-W.; Yang, Y.H.; Kim, H.-W.; Bhang, S.-Y.; Hong, Y.-C. Phthalates exposure and attention-deficit/hyperactivity disorder in school-age children. *Biol. Psychiatry* **2009**, *66*, 958–963. [CrossRef]
15. Ahmad, S.; Zhu, X.; Wang, Q.; Wei, X.; Zhang, S. Microwave-assisted hydrothermal treatment of soybean residue and chitosan: Characterization of hydrochars and role of N and P transformation for Pb (II) removal. *J. Anal. Appl. Pyrolysis* **2021**, *160*, 105330. [CrossRef]
16. Shan, D.; Deng, S.; Zhao, T.; Wang, B.; Wang, Y.; Huang, J.; Yu, G.; Winglee, J.; Wiesner, M.R. Preparation of ultrafine magnetic biochar and activated carbon for pharmaceutical adsorption and subsequent degradation by ball milling. *J. Hazard. Mater.* **2016**, *305*, 156–163. [CrossRef]
17. Plakas, K.V.; Karabelas, A.J. Removal of pesticides from water by NF and RO membranes—A review. *Desalination* **2012**, *287*, 255–265. [CrossRef]
18. Pervez, M.; Balakrishnan, M.; Hasan, S.W.; Choo, K.-H.; Zhao, Y.; Cai, Y.; Zarra, T.; Belgiorno, V.; Naddeo, V. A critical review on nanomaterials membrane bioreactor (NMs-MBR) for wastewater treatment. *NPJ Clean Water* **2020**, *3*, 43. [CrossRef]
19. Kumari, M.; Pulimi, M. Phthalate esters: Occurrence, toxicity, bioremediation, and advanced oxidation processes. *Water Sci. Technol.* **2023**, *87*, 2090–2115. [CrossRef]
20. Miklos, D.B.; Remy, C.; Jekel, M.; Linden, K.G.; Drewes, J.E.; Hübner, U. Evaluation of advanced oxidation processes for water and wastewater treatment–A critical review. *Water Res.* **2018**, *139*, 118–131. [CrossRef]
21. Pidgeon, N.; Porritt, J.; Ryan, J.; Seaton, A.; Tendler, S.; Welland, M.; Whatmore, R. Nanoscience and nanotechnologies: Opportunities and uncertainties. *R. Soc. R. Acad. Eng.* **2004**, *29*, 2004.
22. Bala, S.; Garg, D.; Thirumalesh, B.V.; Sharma, M.; Sridhar, K.; Inbaraj, B.S.; Tripathi, M. Recent strategies for bioremediation of emerging pollutants: A review for a green and sustainable environment. *Toxics* **2022**, *10*, 484. [CrossRef]
23. Roy, A.; Bhattacharya, J. *Nanotechnology in Industrial Wastewater Treatment*; IWA Publishing: London, UK, 2015.
24. Ahmad, S.; Liu, X.; Tang, J.; Zhang, S. Biochar-supported nanosized zero-valent iron (nZVI/BC) composites for removal of nitro and chlorinated contaminants. *Chem. Eng. J.* **2021**, *431*, 133187. [CrossRef]
25. Zhang, S.; Jin, Y.; Chen, W.; Wang, J.; Wang, Y.; Ren, H. Artificial intelligence in wastewater treatment: A data-driven analysis of status and trends. *Chemosphere* **2023**, *336*, 139163. [CrossRef]
26. Pollutants, P.O. Stockholm Convention on Persistent Organic Pollutants. In Proceedings of the Persistent Organic Pollutants Review Committee Twentieth Meeting, Rome, Italy, 23–27 September 2024.
27. Agreement, P. Paris agreement. In Proceedings of the Report of the Conference of the Parties to the United Nations Framework Convention on Climate Change (21st Session, 2015: Paris), Paris, France, 30 November–13 December 2015; p. 2.

water

Article

PFOA and PFOS Pollution in Surface Waters and Surface Water Fish

Bahar Ikizoglu

School of Engineering and Natural Sciences, Environmental Engineering Department, Suleyman Demirel University, Bati Campus, Cunur, Isparta 32260, Turkey; baharikizoglu@sdu.edu.tr; Tel.: +90-246-211-12-81

Abstract: Perfluoroalkyl and poly-fluoroalkyl substances (PFAS) are among the synthetic chemicals employed by various industries since the 1950s and the most critical persistent organic pollutants (POPs) that led to emerging concerns due to high persistency, toxicity, mobility, and environmental bioaccumulation. Although there are more than 5000 types of PFASs, perfluorooctanoic acid (PFOA) and perfluorosulfonic acid (PFOS) are the two chemicals whose employment is highly restricted and banned by the Stockholm Convention. In the present study, certain water resources in the Marmara Region, the most densely populated and industrial region in Turkey, and the waters of Turkey's two largest drinking water reserves, Beyşehir and Eğirdir lakes, were investigated. The study was carried out in two seasons, spring and autumn. The lowest and highest PFOA concentrations were determined between 1.77 ± 0.1 and 6.71 ± 2.9 ng/L in all surface waters, and the highest PFOS concentrations were between <LOQ and 3.27 ng/L. PFOA concentrations were higher when compared to PFOS concentrations in all water sources, and PFOA and PFOS concentrations were lower in spring compared to autumn. In some commercially procured fish from water resources, 7.48 ng/g PFOS was detected in Küçükçekmece Lake pike, and 2.5 ng/g PFOA was identified in Eğirdir Lake trout. PFOA and PFOS were not detected in other fish tissues.

Keywords: PFOA; PFOS; seasonal variations; surface water resources; fish species

1. Introduction

Per- and poly-fluoroalkyl substance (PFAS) compounds have been produced or employed in various industrial processes since the 1950s [1,2]. PFAS chemicals possess unique surface-active properties, are water and oil-repellent, and are highly resistant to heat and acidity. They are widely employed in industries and consumer products due to these properties [3,4]. PFAS are employed as fire protection products, surfactants, surface protectors, food packaging, and consumer products. Since 2000, they have been investigated for potential environmental hazards due to their persistence and bio-accumulative and toxic properties [5]. These properties led to significant concerns about the use of these contaminants of emerging concerns (CECs) [6–8]. Thus, various regulations have been adopted to partially and gradually restrict the use of PFAS. Recently, awareness has been further raised about the negative effects of PFAS exposure on health. Four Northern European EU nations asked the European Commission to develop precautions to reduce PFAS emissions after an earlier call by the European Council on the European Commission to develop an action plan to eliminate all nonessential PFAS emissions. Recently, the European Food Safety Authority (EFSA) recommended a reduction of the tolerable weekly food intake standard. The Dutch National Institute for Public Health and Environment recommended a lower drinking water limit since human PFAS intake depends on drinking water and food consumption. Thus, not only knowledge of the presence of PFAS but also the environmental behavior of PFAS is important to understand present and future environmental and health risks [9].

PFAS assignment is based on the presence of a fully (per) or partially fluorinated (poly)alkyl chain attached to functional groups. PFAS could be classified as polymeric

PFAS (fluoropolymers, side-chain fluorinated polymers, and perfluoropolyether), and non-polymeric PFAS (fluorotelomer (FT), perfluoroalkyl acids (PFAAs), per- and poly-fluoroalkyl ethers (PFPE)). Several studies have been conducted on PFAS; however, these substances have not been fully elucidated and described due to the lack of accurate, specific, and sensitive analytical standards or protocols [10]. Most global research has been conducted on two PFASs, perfluorooctanoic acid (PFOA) and perfluoro-sulfonic acid (PFOS). PFOA and PFOS are both long-chain perfluoro-carboxylic acids with eight or more carbon atoms or perfluoro-sulfonic acids with six or more carbon atoms. PFOA and PFOS are eight-carbon compounds (C8) [11]. PFOA and PFOS have been identified in surface water, groundwater, drinking and coastal waters, landfill leachates, sediments, sludge, and soil between ng/L and μg/L, or ng/g and μg/g. Toxicological studies revealed that PFOA and PFOS could lead to acute or chronic toxicity in freshwater fish, invertebrates, and liver and pancreatic tumors in laboratory animals. Due to the abovementioned factors, PFOA and PFOS were inevitably designated as significant environmental contaminants globally [12]. The US Environmental Protection Agency (USEPA) classified PFOA and PFOS as potential drinking water contaminants in 2009. The US EPA PFOS and PFOA drinking water standard is 4 ng/L [4,13,14].

Since these substances are highly soluble in water, PFOS and PFOA have been reported in most aqueous environments. In contrast with other typical POPs, PFOA and PFOS are highly mobile in aqueous systems due to their ionic nature, high solubility, and negligible vapor pressure when dissolved in water; thus, these compounds could be transferred from commercial, industrial, or domestic discharge to natural waters [15]. The present study aimed to determine PFOA and PFOS concentrations in Beyşehir and Eğirdir lakes—Turkey's largest drinking water sources. It also aimed to determine the concentrations of these pollutants in the Marmara region surface waters, a region with the densest population and the most significant industrial development in Turkey. This study aimed to determine PFOA and PFOS concentrations in Beyşehir and Eğirdir lakes, the largest drinking water sources in Turkey. PFOS and PFOA samples were collected in autumn (October 2022) and spring (April 2023). PFOA and PFOS concentrations were also investigated in fish species procured from fish markets around the water sources.

2. Material and Method

2.1. Standards and Chemicals

Perfluoro-octane sulfonate (PFOS, 98%) was procured from Fluka Analytical (Atlanta, GA, USA), and perfluorooctanoic acid (PFOA, 96%) was procured from Alfa Aesar (Haveril, MA, USA). High-performance liquid chromatography (HPLC) grade methanol (>99.9%) was procured from Sigma-Aldrich (Taufkirchen, Germany), and ammonium acetate and Milli-Q water were procured from Merck Millipore (Darmstadt, Germany). Contaminated water samples were prepared in the laboratory with 1000 mg/L PFOA and PFOS stock solutions in pure methanol. PFOS-PFOA mixtures were prepared by diluting stock solutions with MeOH. The samples were pretreated with Oasis Wax SPE columns in 6 cc cartridges and 150 mg and 30 μm Waters (Milford, MA, USA).

2.2. Sample Collection, Pre-Treatment, and Extraction Procedures

Samples collected into 2 L polypropylene bottles from each surface water source were pairwise analyzed. In the present study, the samples were collected from the Sakarya and Ergene rivers, Terkos, Küçükçekmece, and Büyükçekmece lakes in Marmara Region, Eğirdir Lake in Mediterranean Region, and Beyşehir Lake in Central Anatolia Region the largest drinking water reserve in Turkey. Water and fish samples were transferred to the laboratory in cold chains and stored at −18 °C before the analyses. The sampling locations are presented in Figure 1.

Figure 1. Sample collection locations: (1) Ergene River, (2) Küçükçekmece Lake, (3) Terkos (Durusu) Lake, (4) Büyükçekmece Lake, (5) Sakarya River, (6) Eğirdir Lake, and (7) Beyşehir Lake.

Water samples were sieved through a 1 μm filter and filtered through a 0.45 μm nylon membrane filter (Whatman, UK) before SPE. Samples were collected and analyzed with the "Water Quality-Perfluoro-octane-sulfonate (PFOS) and Perfluoro-octanoate (PFOA) Determination Method for Unfiltered Samples with Solid Phase Extraction and Liquid Chromatography/Mass Spectrometer" as specified in international standards (BS ISO 25101-2009) [16]. Oasis WAX 6 mL Vac cartridges were installed in the Supelco Visiprep vacuum manifold. Cartridges were conditioned sequentially with 4 mL ammonia/methanol solution, 4 mL methanol, and 4 mL ultrapure water. The 500 mL samples and then 4 mL acetate buffer solution were passed through the cartridges, and eluate was discarded. The cartridges were pressurized for 2 more minutes to ensure the removal of all residue. Clean polypropylene tubes were placed in the vacuum manifold. An 8 mL eluate, obtained with 4 mL methanol and 4 mL 0.1% ammonia/methanol solution, was evaporated with nitrogen. The eluate dissolved with 0.5 mL MeOH was shaken for 2 min with a multi-vortex. The product was filtered through a nylon syringe with a 0.22 μm pore diameter and transferred into autosampler vials that included 200 μL inserts for injection [16].

For biota samples, the method developed by Ciccotelli et al. [17] was employed. Briefly, the steps of the analysis were as follows: 5 mL hexane was added to a 2 g fish tissue sample vortexed for 10 min and centrifuged at 4500 rpm for 10 min. The organic phase was discarded, and 10 mL acetonitrile was added; then, the samples were further vortexed for 5 min and centrifuged at 4500 rpm for 10 min. 2 mL supernatant was added to a flask that included 150 mg Al_2O_3, and the product was vortexed for 2 min, centrifuged at 4500 rpm for 10 min and dried with nitrogen. The residue was redissolved with 250 μL MeOH, filtered through a 0.22 μm syringe filter, and vialed. The samples were transferred to LC-MS/MS to determine the concentrations.

2.3. *Ultra-High-Performance (QTRAP® LC-MS/MS) Analysis*

The ABSciex Exigent Expert Ultra LC 100 ABSciex 3200 Q-Trap (MS/MS) chromatography system with a C18 colon, Waters Acuity UPLC BEH, 1.7 µm, 50 × 2.1 mm were employed for chromatographic separation. Mobile phase A included ultrapure water and 20 mM $C_2H_7NO_2$, and mobile phase B included CH_3CN and 0.2% HCOOH. 20 µL sample was injected at a 0.5 mL/min flow rate. The initial mobile phase (60% A, 40% B) was kept for 2 min and then ramped to 90% B for 2 min. The mobile phase was converted to the initial status (60% A) in 4 min. The analysis lasted for 8 min. Extracted ion chromatograms that reflected the PFOA and PFOS retention times are presented in Figure 2.

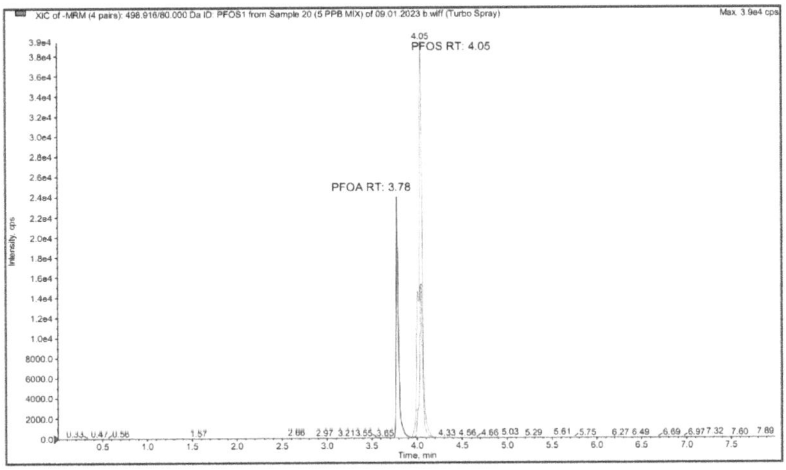

Figure 2. Extracted ion chromatograms for PFOA and PFOS mix reference standards.

3. Findings and Discussion

3.1. *Methodological Performance*

The limit of detection (LOD) and limit of quantification (LOQ) were 50 ng/L and 250 ng/L for PFOA, and 10 ng/L and 50 ng/L for PFOS in water samples, respectively. To determine the recovery rate, Milli Q water was spiked at two concentrations by the standards. The recovery rate was calculated as 85% for PFOA and 89% for PFOS.

Spinner fish *(Vimba Vimba Tenella)* was used as the indicator species, and PFOA and PFOS compounds were not identified as indicator species. Various concentrations of the standard were added to the biota, a matrix-matched calibration curve was plotted with the extracted samples, and quantitative analyses were conducted. LOQ was determined as 2.5 ng/g for PFOA and 1 ng/g for PFOS in biota (fish tissue) samples. MS/MS method parameters are presented in Table 1.

Table 1. MS/MS method parameters for Perfluoroalkyl compounds.

PFC Molecule	Mass Q1 (Da)	Mass Q3 (Da)	DP (Volt)	EP (Volt)	CEP (Volt)	CE (Volt)	CXP (Volt)
PFOA 1	412.852	169	−25	−4.5	−30	−24	−2
PFOA 2	412.852	169	−25	−4.5	−30	−12	−4
PFOS 1	498.916	80	−80	−10.5	−20	−74	0
PFOS 2	498.916	98.9	−80	−10.5	−20	−62	0

$[M\text{-}H]^-$ ions were employed as the precursor ion in tandem mass spectrometry. The two selected reaction monitoring (SRM) transitions between the precursor and the two most prevalent fragment ions were monitored for each compound. The first transition was employed for quantification, and the second was employed to confirm the identity of the target compounds. Further identification criteria in addition to the SRM transition monitoring were also employed for quantification: (i) the UHPLC retention period of the standard compound and the sample compounds were matched, and (ii) the relative content of the two selected analyte SRM transitions was compared between the sample and the standards [18].

Flow injection analysis was employed to determine the source-dependent parameter settings: ESI (-), curtain gas (CUR), 30 V, nitrogen collision gas (CAD) medium, source temperature (TEM) 450 °C, ion spray voltage 4500 V, and ion source gases GS1 and GS2 at 50 and 50 PSIG, respectively.

In the analysis, LC-MS system tubes were replaced with metal ones to prevent instrumental contamination. Furthermore, a second analytical column was added before the injector to prevent contamination from the system or solvents. PFCs were purged from all sample vials, connection elements, and other equipment [19]. All study findings include recovery figures as presented.

3.2. PFOA and PFOS in Surface Waters

The results of the analyses conducted on the samples collected from Ergene and Sakarya rivers, Küçükçekmece, Büyükçekmece, Terkos, Eğirdir, and Beyşehir lakes between October 2022 and April 2023 are presented in Table 2. PFOA and PFOS samples were not collected during winter since PFOA and PFOS concentrations could be low in autumn due to winter precipitation, and the concentrations could be <LOQ or could not be determined. Also, a study on CECs (contaminant of emerging concern) reported that water was diluted in winter [18,20].

Table 2. PFOA and PFOS concentrations in surface waters.

Surface Water	Sample Point Coordinates		Autumn ng/L		Spring ng/L	
	Latitude (N)	Longitude (E)	PFOA	PFOS	PFOA	PFOS
Ergene River	41°21′31.92	27°81′39.82	3.7 ± 0.3	1.1 ± 0.1	2.7 ± 0.2	1.1 ± 0.1
Sakarya River	40°47′45.18	30°26′13.56	2.5 ± 0.3	1.4 ± 0.2	1.8 ± 0.1	1.0 ± 0.1
Küçükçekmece Lake	41°0′2.61	28°45′53.28	4.2 ± 0.1	3.3 ± 0.2	3.2 ± 0.7	2.2 ± 0.2
Büyükçekmece Lake	41°2′13.92	28°33′40.83	3.4 ± 0.2	1.5 ± 0.2	2.4 ± 0.1	1.5 ± 0.2
Terkos Lake	41°19′19.59	28°37′15.84	5.3 ± 1.1	0.7 ± 0.1	4.9 ± 0.3	0.9 ± 0.2
Eğirdir Lake	37°53′45.71	30°50′45.04	6.7 ± 2.9	2.6 ± 0.8	4.6 ± 0.2	2.3 ± 1.2
Beyşehir Lake	37°54′7.43	31°28′37.66	2.9 ± 0.3	<LOQ	1.8 ± 0.1	<LOQ

Note(s): <LOQ: analysis result < limit of quantification.

As seen in Table 2, PFOA and PFOS were identified in all surface water samples. The highest PFOA concentration (6.1 ± 2.9 ng/L) was observed in Eğirdir Lake in autumn, while the lowest PFOA (1.8 ± 0.1 ng/L) was observed in Sakarya River in spring. The highest PFOS concentration (3.3 ± 0.2 ng/L) was observed in Küçükçekmece Lake in the autumn, and the lowest PFOS concentration (<LOQ) was observed in Beyşehir Lake in the autumn. In the autumn, the season where the highest concentrations were observed, PFOA concentrations were 6.7 ± 2.9 ng/L in Eğirdir Lake, 5.3 ± 1.1 ng/L in Terkos Lake, 4.3 ± 1.1 ng/L in Küçükçekmece Lake, 3.7 ± 0.3 ng/L in Ergene River, 3.4 ± 0.2 ng/L in Büyükçekmece Lake, 2.9 ± 0.3 ng/L in Beyşehir Lake, 2.5 ± 0.3 ng/L in Sakarya River, and PFOS concentrations were 3.3 ± 0.2 ng/L in Küçükçekmece Lake, 2.6 ± 0.8 ng/L in Eğirdir Lake, 1.5 ± 0.2 ng/L in Büyükçekmece Lake, 1.5 ± 0.2 ng/L in Sakarya River, 1.1 ± 0.1 ng/L in Ergene River, 0.7 ± 0.1 ng/L in Terkos Lake, and <LOQ in Beyşehir Lake.

In autumn, the analysis findings were generally higher than in spring. This could be due to the dilution of existing PFOA and PFOS concentrations in the receiving water environments due to winter precipitation. PFOA concentrations were higher in all water sources when compared to PFOS concentrations [20]. Seasonal PFOA and PFOS variations by surface water source are presented in Figure 3.

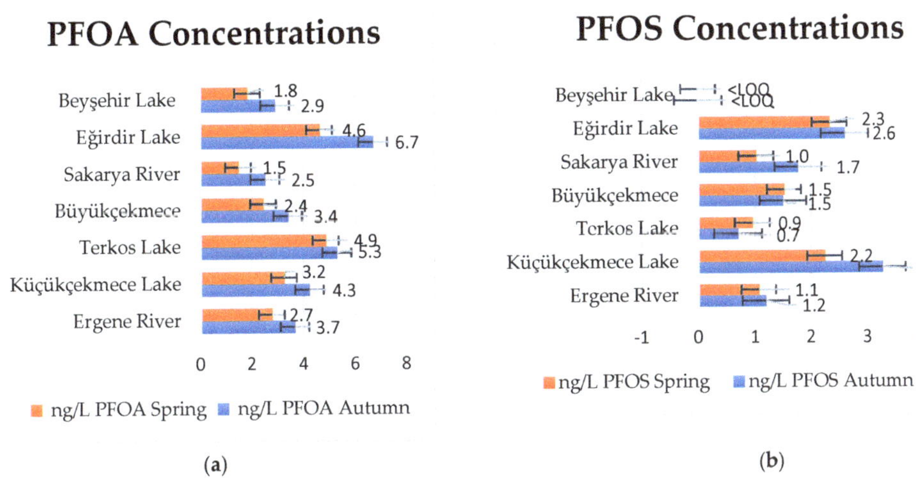

Figure 3. PFOA (**a**) and PFOS (**b**) concentrations in surface water in spring and autumn.

Although the Marmara Region is the most densely populated region with high industrial development in Turkey, the investigated surface water resources exhibited relatively lower PFOA and PFOS concentrations when compared to other countries. PFOA concentrations were 2.91 and 1.94 ng/L, and PFOS concentrations were <LOQ in autumn and spring in Lake Beyşehir, the largest drinking water reserve in Turkey located in the Central Anatolia Region. Lake Eğirdir, located in the Mediterranean Region, PFOA concentrations were 6.71 ± 2.9 and 4.64 ± 0.2 ng/L, and PFOS concentrations were 2.6 ± 0.8 and 2.3 ± 1.2 ng/L in autumn and spring, respectively. PFOA concentrations were higher than PFOS concentrations in surface water (Figure 3). This could be due to the high use of PFOA and the earlier prohibition (deprecation) of PFOS. Drinking water supplies are vulnerable to PFOA and PFOS contamination that originate in several sources. Wastewater treatment plant discharge, biodegradation of precursors during wastewater treatment, industrial discharges, and land disposal of biosolids are potential PFOA and PFOS sources in drinking water reserves [21].

The concentrations identified in the waters of certain rivers and lakes in some countries are presented in Table 3. The present study findings revealed lower concentrations when compared to these figures. Before 2002, US EPA published PFOS and PFOA limits in drinking water as 150 µg/L and 1 µg/L. In 2009, these limits were lowered to 0.2 µg/L for PFOS and 0.4 µg/L for PFOA, and these levels were further reduced to 70 ng/L for PFOS and PFOA in 2016 [4], and the current US EPA PFOS and PFOA drinking water standard is 4 ng/L [14]. Since PFAS toxicity is still under investigation, the 4 ng/L limit could be further reduced in the future. Thus, comprehensive studies are required on PFOA and PFOS, which are toxic, persistent, and endocrine disruptors, especially in surface waters that could be consumed as drinking water [22,23].

Takagi et al. [36] investigated drinking water treatment plants and raw surface water sources that fed these plants and determined that PFOS and PFOA concentrations were 3.7 and 48 ng/L in treatment plant effluent, 1.3 and 15 ng/L in raw river water influent, PFOS and PFOA were 2.2 and 36 ng/L in raw lake water, and <0.50 and 6.5 ng/L in the treatment plant effluent. The drinking water standard determined by the EPA for PFOA and PFOS concentrations on March 2023 was 4 ng/L [14], and various measures should be adopted for global implementation. Particularly, PFASs in drinking water sources could seriously threaten water reservoirs, aquatic ecosystems, and public health. Urgent research is required on PFASs known as forever chemicals, toxicity studies should be conducted on aquatic ecosystems in drinking water reservoirs, PFAS-contaminated water sources should be rehabilitated, and advanced treatment methods such as ion exchange, adsorption (especially granular activated carbon, GAC), nanofiltration, and reverse osmosis (NF, RO) should be included in drinking water treatment facility effluents.

Table 3. PFOA and PFOS concentrations in certain country rivers and lakes.

Country	Water Body	PFOA (ng/L)	PFOS (ng/L)	References
Canada	Lake Ontorio	4.4–44	3.1–37	[24]
	Grand River	6.5–9.4	10.2–20.0	[25]
	Lake Ontorio	<0.25–33	<10	[26]
USA	Hudson River	22–173	1.5–3.4	[27]
	Lake Champlain	10–46	0.8–7.7	
	Lake Erie	21–47	11–39	[28]
China	Hong Lake	12.2	6.0	[29]
	Dong Lake	3.3	3.8	
	Shenzhen River	30.8	10.2	
	Pearl River	6.2–14.3	5.7–14.1	
India	Gangs River	0.033–2.0	0.04–8.4	[30]
Japan	Tsurumi River	13.4–15.9	179.6–179.9	[31]
	Uji River	100–110	8.7–10	[32]
Spain	Catalan Rivers	0.79–9.63	1.09–9.56	[33]
	L'Albufera lake	0.03–10.90	0.10–4.80	[34]
Taiwan	Xioali River	17.3	82	[35]
	Tauchien River	10.9	48.9	
	Keya River	310	5440	

3.3. *PFOA and PFOS in Fish Tissue*

Fish purchased from fish markets around the lakes were analyzed for PFOA and PFOS. Carp biota of the same size and weight were procured from Lake Eğirdir and Beyşehir Lake markets in 2 seasons, pike procured from Küçükçekmece Lake, and trout procured from Eğirdir Lake were available only one season. Fish samples from other water bodies were not available at fish markets at these water bodies and were not analyzed. PFOA and PFOS findings for investigated fish species are presented in Table 4.

Table 4. PFOA and PFOS concentrations in fish tissue samples.

Surface Waters	ng/g	
	PFOA	PFOS
Lake Küçükçekmece Pike (*Esox Lucius*)	<LOQ	7.5 ± 0.3
Lake Eğridir Trout (*Salmo Trutta*)	2.5 ± 0.1	<LOQ
Lake Eğirdir 1st season (*Cyprinus Carpio*)	<LOQ	<LOQ
Lake Eğirdir 2nd season (*Cyprinus Carpio*)	<LOQ	<LOQ
Lake Beysehir Carp 1st season (*Cyprinus Carpio*)	<LOQ	<LOQ
Lake Beyşehir Carp 2nd season (*Cyprinus Carpio*)	<LOQ	<LOQ

Note(s): <LOQ: analysis result < limit of quantification.

PFOA and PFOS concentrations in *Esox Lucius* collected at Küçükçekmece Lake were <LOQ and 7.48 ng/g, respectively, while 2.5 ng/g and <LOQ concentrations were identified in *Salmo Trutta* collected at Eğirdir Lake. PFOA and PFOS were not identified in *Cyprinus Carpio* collected at Lake Eğirdir and *Cyprinus Carpio* collected at Lake Beyşehir in autumn and spring (Table 4). This could be because young biota of the same size and weight (fresh fish weight of about 300 g) were used to compare the two lakes.

While the tolerable daily intake limits previously set by EFSA (Panel on Contaminants in the Food Chain) were 150 ng/kg body weight for PFOS and 1500 ng/kg body weight for PFOA, in 2018, these limits were set at 13 ng/kg body weight and 6 ng/kg body weight per week. The reason for this significant decrease in the limits was the findings reported by toxicological and epidemiological studies, which demonstrated that these compounds could lead to health problems. In 2020, EFSA updated the weekly intake limit for total PFOA, PFOS, PFNA, and PFHxS compounds to 4.4 ng/kg body weight. It was determined that fish contributed 50% to the PFAS intake [37]. Paiano et al. [38] investigated the PFOA and PFOS levels in wild and farmed fish filets collected in the Mediterranean Sea and reported that the mean PFOA and PFOS concentrations were 0.19 ng/g and 1.24 ng/g for 52 wild fish species, and <0.05 ng/g and 0.05 ng/g for 13 farmed fish species, respectively. Squadrone et al. [39] studied 40 European sea bass and 50 European whitefish samples collected at Lake Maggiore in the Italian Alps and found that the highest PFOS level was 45.8 ng/g in female sea bass.

In a study that investigated PFOS, PFOA, PFNA, and PFHxS concentrations in 24 fish species and seafood in Greece, it was reported that the total PFAS concentration was between <LOQ and 20.4 ng/g, and the highest concentration was identified in butternut fish. While PFOS was identified in 50% of the analyzed samples, PFNA and PFOA were detected only in 10%, and PFHxS was not detected in any [40]. Fifty fish samples of six species in different ecosystems were screened for nine PFAS compounds, and PFOS, PFOA, and PFOSA were identified in all samples. The highest PFOS concentration was (12.83 μg/kg) in tabby fish. The mean PFOS concentration was between 0.3 and 750 μg/kg in 65 fish samples (mullet, perch, long perch) collected at a lake far from large urban centers and any pollutant source, and PFOA, PFNA, and PFHxS concentrations were below the detection limit [41]. Commercial fish are typically bred in marine fisheries. Marine fish species generally contain significantly lower PFAS concentrations when compared to freshwater fish. Thus, freshwater fish consumers are particularly at risk. Alarming levels of PFAS bioaccumulation in fish were observed even when water PFAS levels were quite low or even undetectable. This could be because instantaneous surface water samples may not represent the level that the fish were exposed to, or bioaccumulation in fish could be higher than the concentration in the water [42].

PFAS (per- and poly-fluoroalkyl substances) are considered among critical pollutant groups due to their adverse effects on human health after exposure. The most common sources of PFAS exposure include the consumption of contaminated drinking water, certain fish and shellfish, food packages that contain PFAS, transfer from stain-resistant carpets and textiles, and direct contact with industrial products that contain PFAS. Although PFAS production started in the 1940s, the sources and toxicity of these substances were initially studied in the 1990s. Due to their toxicity potential, mobility, stability (resistance to degradation in the environment), and bioaccumulation properties, PFASs are highly hazardous and often referred to as "forever chemicals" [6]. The impact of chronic exposure to low PFAS concentrations or short-term exposure to high PFAS concentrations are still not fully elucidated. When surface waters are contaminated by various substances, the consumption of PFAS-contaminated surface water through ingestion or skin contact could lead to adverse health outcomes. Health issues associated with PFAS exposure that were identified in infants, children, adults, pregnant women, and the elderly include cancer, diabetes, high blood pressure, asthma, and weak immune system. Reported health problems include kidney and testicular cancers, impaired liver function, chronic kidney damage, cardiovascular diseases, inflammatory bowel diseases, elevated cholesterol levels, osteoarthritis, thyroid and other hormonal disorders, delayed breast development, reduced fetal growth, hypertension or preeclampsia in pregnancy, high miscarriage risk, preterm birth, low birth weight, childhood obesity, emotional and behavioral disorders, and early or delayed puberty. Due to these negative health effects associated with PFAS exposure, further research could be recommended on the effects of both short-term and long-term exposure [43,44].

Research on PFASs in Turkey was conducted on PTFE-coated non-stick containers [45], food and beverages [46], and bottled and drinking water, among others [46,47]. Endirlik et al. [47] reported that the maximum PFOA concentration was 2.37 ng/L, and PFOS concentration was 1.93 ng/L in drinking water, and PFOA was 0.1 ng/L, and could not detect PFOS in bottled water in Turkey. In Turkey, PFAS contaminants should further be investigated in various environments such as food, water sources, water and wastewater treatment plants, and soil around industrial areas to determine PFAS risk levels.

4. Discussion

The investigation of PFOA and PFOS contamination in surface waters and fish allows the analysis of results. This study provides a framework within which the findings of the determination of PFOA and PFOS levels in surface waters and surface water fish can be considered. PFOA between 1.77 $\pm$ 0.1 and 6.71 $\pm$ 2.9 ng/L and PFOS between 0.7 $\pm$ 0.1 ng/L and 3.27 $\pm$ 0.2 ng/L were detected in the indicated surface water sources. In Lake Eğirdir trout tissue (*Salmo Trutta*) 2.5 ng/g PFOA and in Lake Küçükçekmece pike (*Esox Lucius*) tissue 7.48 ng/g PFOS were detected. The higher concentration of PFOS in *Esox Lucius* than in Kücükcekmece Lake water may be explained by the fact that the instantaneous sample taken may not be representative of the level to which the fish is exposed or that bioaccumulation in the fish occurs at a higher degree than the concentration in the water. Our findings warrant further reflection, especially on the ecological impacts of these pollutants. Some studies have shown that these pollutants, which are found at lower concentrations in surface waters, can reach higher concentrations in drinking water treatment plant effluents, while others have shown that these concentrations are lower. In particular, seasonal analyses of surface water intakes of drinking water treatment plants and various stages of drinking water treatment plants could be investigated in future studies. Despite their widespread use, PFOA and PFOS are currently restricted and banned due to their persistence, mobility, toxicity, and bioaccumulation in the food chain. Furthermore, PFOA and PFOS are the most prevalent pollutants in the environment, and the EPA was determined as 4 ng/L in drinking water in 2023. These pollutants reach surface waters through various routes such as atmospheric transport, industrial discharges, wastewater discharges, and surface runoff, pollute rivers and lakes, disrupt aquatic ecosystems, and also pose a serious global threat to all living organisms due to bioaccumulation, persistence, mobility and toxicity in the food chain. Especially water resources used as drinking water reservoirs and river waters feeding these resources should be regularly monitored for PFAS chemicals. Detection and seasonal monitoring of these chemicals, which pose a significant threat to human health, in surface water sources and drinking water treatment plants should be a requirement in every country. New processes (NF: nanofiltration, RO: reverse osmosis, etc.) should be adopted as the final treatment step in WWTPs in order to prevent the access of PFOA and PFOS to surface water sources and aquatic ecosystems, especially from wastewater treatment plants. In this context, new information should be obtained to guide decision-makers in the development of stricter regulations to control PFOA and PFOS pollution. Finally, it is important that the findings presented in this paper be taken into account by legislators in order to protect water resources and improve human health.

Funding: This study was supported by SDU—Süleyman Demirel University Scientific Research Projects Coordination Department with project number FAB-2022-8671.

Institutional Review Board Statement: Not applicable.

Informed Consent Statement: Not applicable.

Data Availability Statement: To access the data used in the creation of this paper, contact the corresponding author.

Acknowledgments: I would like to thank my family for their support and encouragement in this study, as always.

Conflicts of Interest: The author declares no conflicts of interest.

References

1. Fuertes, I.; Gomez-Lavin, S.; Elizalde, M.P.; Urtiaga, A. Perfluorinated alkyl substances (PFASs) in northern Spain municipal solid waste landfill leachates. *Chemosphere* **2017**, *168*, 399–407. [CrossRef]
2. Blake, B.E.; Fenton, S.E. Early life exposure to per- and polyfluoroalkyl substances (PFAS) and latent health outcomes: A review including the placenta as a target tissue and possible driver of peri- and postnatal effects. *Toxicology* **2020**, *443*, 152565. [CrossRef]
3. Glüge, J.; Scheringer, M.; Cousins, I.T.; DeWitt, J.C.; Goldenman, G.; Herzke, D.; Lohmann, R.; Ng, C.A.; Trier, X.; Wang, Z. An overview of the uses of per- and polyfluoroalkyl substances (PFAS). *Environ. Sci. Process. Impacts* **2020**, *22*, 2345–2373. [CrossRef]
4. Cordner, A.; Goldenman, G.; Birnbaum, L.S.; Brown, P.; Miller, M.F.; Mueller, R.; Patton, S.; Salvatore, D.H.; Leonardo, L. The true cost of PFAS and the benefits of acting now. *Environ. Sci. Technol.* **2021**, *55*, 9630–9633. [CrossRef]
5. Cousins, I.T.; DeWitt, J.C.; Glüge, J.; Goldenman, G.; Herzke, D.; Lohmann, R.; Miller, M.; Ng, C.A.; Scheringer, M.; Vierke, L.; et al. Strategies for group in per-and polyfluoroalkyl substances (PFAS) to protect human and environmental health. *Environ. Sci. Process. Impacts* **2020**, *22*, 1444–1460. [CrossRef]
6. Sunderland, E.M.; Hu, X.C.; Dassuncao, C.; Tokranov, A.K.; Wagner, C.C.; Allen, J.G. A review of the pathways of human exposure to poly- and perfluoroalkyl substances (PFASs) and present understanding of health effects. *J. Expo. Sci. Environ. Epidemiol.* **2019**, *29*, 131–147. [CrossRef]
7. Lee, Y.M.; Lee, J.Y.; Kim, M.K.; Yang, H.; Lee, J.E.; Son, Y.; Kho, Y.; Choi, K.; Zoh, K.D. Concentration and distribution of per- and polyfluoroalkyl substances (PFAS) in the Asan Lake area of South Korea. *J. Hazard. Mater.* **2020**, *381*, 120909. [CrossRef]
8. Gaines, L.G.T. Historical and current usage of per-and polyfluoroalkylsubstances (PFAS): A literature review. *Am. J. Ind. Med.* **2023**, *66*, 353–378. [CrossRef]
9. Gerardu, T.; Dijkstra, J.; Beeltje, H.; Duivenbode, A.V.; Griffioen, J. Accumulation and transport of atmospherically deposited PFOA and PFOS in undisturbed soils downwind from a fluoropolymers factory. *Environ. Adv.* **2023**, *11*, 100332. [CrossRef]
10. Podder, A.; Anwar-Sadmani, A.H.M.; Reinhart, D.; Chang, N.B.; Goel, R. Per and poly-fluoroalkyl substances (PFAS) as a contaminant of emerging concern in surface water: A transboundary review of their occurrences and toxicity effects. *J. Hazard. Mater.* **2021**, *419*, 126361. [CrossRef]
11. Pontus, F. Regulation of Perfluorooctanoic Acid (PFOA) and Perfluorooctane Sulfonic Acid (PFOS) in Drinking Water: A Comprehensive Review. *Water* **2019**, *11*, 2003. [CrossRef]
12. Li, Y.; Liu, Y.; Shi, G., Liu, C.; Hao, Q.; Wu, L. Occurrence and Risk Assessment of Perfluorooctanoate (PFOA) and Perfluorooctane Sulfonate (PFOS) in Surface Water, Groundwater and Sediments of the Jin River Basin, Southeastern China. *Bull. Environ. Contam. Toxicol.* **2022**, *108*, 1026–1032. [CrossRef]
13. Vo, P.H.N.; Ngo, H.H.; Guo, W.; Nguyen, T.M.H.; Li, J.; Liang, H.; Deng, L.; Chen, Z.; Nguyen, T.A.H. Poly-and perfluoroalkyl substances in water and wastewater: A comprehensive review from sources to remediation. *J. Water Process Eng.* **2020**, *36*, 101393. [CrossRef]
14. U.S Environmental Protection Agency. Proposed PFAS National Primary Drinking Water Regulation. Available online: https://www.epa.gov/sdwa/and-polyfluoroalkyl-substances-pfas (accessed on 10 August 2024).
15. Flores, C.; Ventura, F.; Alonso, J.M.; Caixach, J. Occurrence of perfluorooctane sulfonate (PFOS) and perfluorooctanoate (PFOA) in N.E. Spanish surface waters and their removal in a drinking water treatment plant that combines conventional and advanced treatments in parallel lines. *Sci. Total Environ.* **2013**, *461–462*, 618–626. [CrossRef]
16. ISO 25101. 2009 Water Quality. Available online: https://www.iso.org/standard/42742.html (accessed on 29 October 2023).
17. Ciccotelli, V.; Abete, C.M.; Stefania Squadrone, S. PFOS and PFOA in cereals and fish: Development and validation of a high-performance liquid chromatography-tandem mass spectrometry method. *Food Control* **2016**, *59*, 46–52. [CrossRef]
18. Ikizoglu, B.; Turkdogan, F.I.; Kanat, G.; Aydıner, C. Seasonal analysis of commonly prescribed antibiotics in Istanbul city. *Environ. Monit. Assess.* **2023**, *195*, 566. [CrossRef]
19. Llorca, M.; Farré, M.; Tavano, M.S.; Alonso, B.; Koremblitn, G.; Barceló, D. Fate of a broad spectrum of perfluorinated compounds in soils and biota from Tierra del Fuego Antarctica. *Environ. Pollut.* **2012**, *163*, 158–166. [CrossRef]
20. Ozvardarlı, A. Quantitative Determination of Perfluorooctane Sulfonate (PFOS) and Perfluorooctanoic Acid (PFOA) Compounds in Surface Waters in Ergene Waterbasin and Evaluation of the Treatability of These Compounds. Ph.D. Thesis, Namık Kemal University, Graduate School of Natural and Applied Sciences, Tekirdağ, Türkiye, 2020.
21. Huset, C.A.; Chiaia, A.C.; Barofsky, D.F.; Jonkers, N.; Kohler, H.P.E.; Ort, C.; Giger, W.; Field, J.A. Occurrence and mass flows of fluorochemicals in the Glatt valley watershed, Switzerland. *Environ. Sci. Technol.* **2008**, *42*, 6369–6377. [CrossRef]
22. Liu, C.; Gin, K.Y.H.; Chang, V.W.C.; Goh, B.P.L. Novel perspectives on the bioaccumulation of PFCs–The concentration dependency. *Environ. Sci. Technol.* **2011**, *45*, 9758–9764. [CrossRef] [PubMed]
23. Zhu, H.; Kannan, K. Distribution and partitioning of perfluoroalkyl carboxylic acids in surface soil, plants, and earthworms at a contaminated site. *Sci. Total Environ.* **2019**, *647*, 647–961. [CrossRef]
24. Awad, E.; Xianming Zhang, X.; Bhavsar, S.P.; Steve Petro, S.; Patrick, W.; Crozier, P.W.; Reiner, E.J.; Fletcher, R.; Tittlemier, S.A.; Braekevelt, E. Long-Term Environmental Fate of Perfluorinated Compounds after Accidental Release at Toronto Airport. *Environ. Sci. Technol.* **2011**, *45*, 8081–8089. [CrossRef] [PubMed]
25. Furdui, V.I.; Crozier, P.W.; Reiner, E.J.; Mabury, S.A. Trace level determination of perfluorinated compounds in water by direct injection. *Chemosphere* **2008**, *73*, 524–530. [CrossRef]

26. Moody, C.A.; Martin, J.W.; Kwan, W.C.; Derek, C.G.; Muir, D.G.C.; Mabury, S. Monitoring Perfluorinated Surfactants in Biota and Surface Water Samples Following an Accidental Release of Fire-Fighting Foam into Etobicoke Creek. *Environ. Sci. Technol.* **2002**, *36*, 545–555. [CrossRef]
27. Sinclair, E.; Mayack, D.T.; Kenneth Roblee, K.; Nobuyoshi Yamashita, N.; Kannan, K. Occurrence of Perfluoroalkyl Surfactants in Water, Fish, and Birds from New York State. *Arch. Environ. Contam. Toxicol.* **2006**, *50*, 398–410. [CrossRef] [PubMed]
28. Boulanger, B.; Wargo, J.; Schnoor, J.S.; Hornbuckle, K.C. Detection of Perfluorooctane Surfactants in Great Lakes Water. *Environ. Sci. Technol.* **2004**, *38*, 4064–4070. [CrossRef] [PubMed]
29. Jin, Y.H.; Liu, W.; Sato, I.; Nakayama, S.F.; Sasaki, K.; Saito, N.; Tsuda, S. PFOS and PFOA in environmental and tap water in China. *Chemosphere* **2009**, *77*, 605–611. [CrossRef] [PubMed]
30. Yeung, L.W.Y.; Yamashita, Y.; Taniyasu, N.; Lam, P.K.S.; Sinha, R.K.; Borole, D.V.; Kannan, K. A survey of perfluorinated compounds in surface water and biota including dolphins from the Ganges River and in other waterbodies in India. *Chemosphere* **2009**, *76*, 55–62. [CrossRef] [PubMed]
31. Zushi, Y.; Takeda, T.; Masunaga, S. Existence of nonpoint source of perfluorinated compounds and their loads in the Tsurumi River basin, Japan. *Chemosphere* **2008**, *71*, 1566–1573. [CrossRef]
32. Senthilkumar, K.; Ohi, E.; Sajwan, K.; Takasuga, T.; Kannan, K. Perfluorinated compounds in river water, river sediment, market fish, and wildlife samples from Japan. *Bull. Environ. Contam. Toxicol.* **2007**, *79*, 427–431. [CrossRef]
33. Sánchez-Avila, J.; Meyer, J.; Lacorte, S. Spatial distribution and sources of perfluorochemicals in the NW Mediterranean coastal waters (Catalonia, Spain). *Environ. Pollut.* **2010**, *158*, 2833–2840. [CrossRef]
34. Pico, Y.; Blasco, C.; Farré, M.; Barceló, D. Occurrence of perfluorinated compounds in water and sediment of L'Albufera Natural Park (València, Spain). *Environ. Sci. Pollut. Res.* **2012**, *19*, 946–957. [CrossRef]
35. Lin, A.Y.C.; Panchangam, S.C.; Lo, C.C. The impact of semiconductor, electronics and optoelectronic industries on downstream perfluorinated chemical contamination in Taiwan rivers. *Environ. Pollut.* **2009**, *157*, 1365–1372. [CrossRef]
36. Takagi, S.; Adachi, F.; Miyano, K.; Koizumi, Y.; Tanaka, H.; Watanabe, I.; Tanabe, S.; Kannan, K. Fate of perfluorooctanesulfonate and perfluorooctanoate in drinking water treatment processes. *Water Res.* **2011**, *45*, 3925–3932. [CrossRef] [PubMed]
37. EFSA Panel on Contaminants in the Food Chain (EFSA CONTAM Panel). Risk to human health related to the presence of perfluoroalkyl substances in food. *EFSA J.* **2020**, *18*, 6223. [CrossRef]
38. Paiano, V.; Fattore, E.; Carra, A.; Generoso, C.; Fanelli, R.; Bagnati, R. Liquid chromatography–tandem mass spectrometry analysis of perfluorooctane sulfonate and perfluorooctanoic acid in fish fillet samples. *J. Anal. Methods Chem.* **2012**, 719010. [CrossRef]
39. Squadrone, S.; Ciccotelli, V.; Favoro, L.; Scanzio, T.; Prearo, M.; Abete, M.C. Fish consumption as a source of human exposure to perfluorinated alkyl substances in Italy: Analysis of two edible fish from Lake Maggiore. *Chemosphere* **2014**, *114*, 181–186. [CrossRef]
40. Costopoulou, D.; Vassiliadou, I.; Leondiadis, L. PFASs intake from fish, eggs and drinking water in Greece in relation to the safety limits for weekly intake proposed in the EFSA scientific opinion of 2020. *Chemosphere* **2022**, *286*, 131851. [CrossRef]
41. Augustsson, T.; Lennqvist, C.M.G.; Osbeck, P.; Tibblin, A.; Glynn, M.A.; Nguyen, E.; Westberg, R. Consumption of freshwater fish: A variable but significant risk factor for PFOS exposure. *Environ. Res.* **2021**, *192*, 110284. [CrossRef]
42. Goodrow, S.M.; Ruppel, B.; Lippincott, R.L.; Post, G.B.; Procopio, N.A. Investigation of levels of perfluoroalkyl substances in surface water, sediment and fish tissue in New Jersey, USA. *Sci. Total Environ.* **2020**, *729*, 138839. [CrossRef]
43. Fenton, S.E.; Ducatman, A.; Boobis, A.; DeWitt, J.C.; Lau, C.; Ng, C.; Smith, J.S.; Roberts, S.M. Per- and Polyfluoroalkyl Substance Toxicity and Human Health Review: Current State of Knowledge and Strategies for Informing Future Research. *Environ. Toxicol. Chem.* **2020**, *40*, 606–630. [CrossRef] [PubMed]
44. Tang, Z.W.; Hamid, F.S.; Yusoff, I.; Chan, V. A review of PFAS research in Asia and occurrence of PFOA and PFOS in groundwater, surface water and coastal water in Asia. *Groundw. Sustain. Dev.* **2023**, *22*, 100947. [CrossRef]
45. Toptancı, I.; Ketenoglu, O.; Kıralan, M. Assessment of the migration of perfuorinated compounds and primary aromatic amines from PTFE-coated non-stick cookware marketed in Turkey. *Environ. Sci. Pollut. Res.* **2022**, *29*, 38535–38549. [CrossRef]
46. Sungur, Ş.; Köroğlu, M.; Turgut, F. Determination of perfluorooctanoic acid (PFOA) and perfluorooctane sulfonic acid (PFOS) in food and beverages. *Int. J. Environ. Anal. Chem.* **2018**, *98*, 360–368. [CrossRef]
47. Ünlü-Endirlik, B.; Bakır, E.; Boşgelmez, İ.İ.; Eken, A.; Narin, İ.; Gürbay, A. Assessment of perfluoroalkyl substances levels in tap and bottled water samples from Turkey. *Chemosphere* **2019**, *235*, 1162–1171. [CrossRef]

water

MDPI

Article

Research Progress on Migratory Water Birds: Indicators of Heavy Metal Pollution in Inland Wetland Resources of Punjab, Pakistan

Shifa Shaffique [1,†], Sang-Mo Kang [1,†], Muhammad Ahsan Ashraf [2], Ali Umar [3], Muhammad Saleem Khan [3], Muhammad Wajid [3], Abdullah Ahmed Al-Ghamdi [4] and In-Jung Lee [1,*]

1 Department of Applied Biosciences, Kyungpook National University, Daegu 41566, Republic of Korea; shifa.2021@knu.ac.kr (S.S.)
2 Department of Zoology, Division of Science and Technology, University of Education, Lahore 54000, Pakistan
3 Department of Zoology, Faculty of Life Sciences, University of Okara, Okara 56130, Pakistan;
4 Department of Botany and Microbiology, College of Science, King Saud University, P.O. Box 2455, Riyadh 11451, Saudi Arabia
* Correspondence: ijlee@knu.ac.kr
† These authors contributed equally to this work.

Abstract: The heavy metal burden on natural freshwater ecosystems is uninterruptedly increasing, which could affect their biodiversity, particularly regarding avian species. Three river barrages were selected for the sampling of water birds from autumn 2021 to spring 2022. Seven heavy metals—nickel (Ni), copper (Cu), cobalt (Co), zinc (Zn), lead (Pb), cadmium (Cd), and manganese (Mn)—were estimated in the hearts and livers of *Fulica atra* (Common Coot), *Anas strepera* (Gad-wall), and *Anas crecca* (Eurasian Teal) (Linnaeus, 1758) by atomic absorption spectrometry. The mean concentrations of metals were found to be significantly ($p < 0.05$) different among the sam-pling sites, species, and tissues. In the livers of *F. atra* and *A. strepera*, respectively, the highest mean concentration among the metals belonged to Zn (521 μg/g), and the lowest belonged to Cd (0.17 μg/g). The concentrations of Zn, Cu, Pb, Cd, and Co were highest in A. strepera heart samples. However, the concentrations of Zn, Pb, Cd, and Mn were found to be highest in the livers of *F. atra*. A comparison between both tissues indicated that the concentration of Ni is significantly ($p < 0.05$) higher (except for *F. atra* from Trimmu barrage) in the livers of water birds and that the concentration of Cd is significantly higher in the hearts of water birds. The mean metal concentrations were higher than the background limits reported in various studies, suggesting that the wintering water birds of Pakistan are under a significant load of heavy metal pollution. Histopathological analysis suggested that the observed heavy metals altered the normal histologies of hearts and livers of *Fulica atra* (Common Coot), *Anas strepera* (Gadwall), and *Anas crecca* (Eurasian Teal) sampled from three wetlands of Punjab Pakistan.

Keywords: atomic absorption spectrometry; heart; liver; water birds

1. Introduction

Natural aquatic bodies receive pollutants, including heavy metals, from natural and artificial sources. The exposure and uptake of metals vary in a species because of differences in age, gender, body size, diet, and trophic level in a food chain. For wetland ecosystems, water birds are considered as the main indicators, being more vulnerable to heavy metals exposure [1]. Exposure to metal contaminants is also influenced by the solubility, bioavailability, mobility, and binding affinity of metals with tissues [2,3]. Migratory water birds use many wetlands of Pakistan as their wintering sites. These wetlands are mainly contaminated with heavy metal population coming from different sources (i.e., untreated urban sewage, industrial effluents, and agricultural wastes) [4].

The wetlands of Punjab have rich avifuana. The water of Chenab River and Jhelum River collects at Trimmu barrage, which was constructed to control floods in Jhang city and ensure the supply of irrigation water. It is a combination of freshwater and terrestrial ecosystems and is home to a wide variety of animals and plants [5]. Trimmu barrage serves as a significant wetland site that attracts migratory and endemic species of birds. The barrage area encompasses a diverse range of aquatic and terrestrial ecosystems, making it a highly suitable habitat for a wide array of migratory and endemic avian species [6]. Zaman et al. [7] observed fifty-eight species of resident and migratory birds at Trimmu barrage. Taunsa barrage and Chashma barrage serve as the significant wetland habitats for both international and local migratory bird species. This includes various types of waterfowl, waders, and avifauna that migrate over long distances.

The present study was conducted to (1) estimate heavy metal concentrations in the wetlands of Punjab, Pakistan; (2) compare heavy metal concentrations between wetlands; and (3) estimate differences in the accumulation of metals among species and tissues.

2. Materials and Methods

2.1. Study Area and Sampling Sites

The present study was conducted on three major wetlands (Chashma barrage, Taunsa barrage, and Trimmu barrage) of Punjab, Pakistan. Chashma barrage was built on the Indus River in 1971. Five lakes with low levels of water arise from different embankments of the barrage. An area of about 250 hectares is occupied by each lake. The peripheral area is leased to farmers for crop cultivation. The water level in all lakes varies with the season (up to 8 m in flood season and 0.2 m in dry period). Taunsa barrage was also constructed on the Indus River. It is a reservoir of water that is mainly used for irrigation. On the barrage, five embankments are formed, which prevents the drying of lakes and maintains shallow waters when there is a shortage of water in the main river channel. The water level in sideway channels and lagoons varies between 0.2 m and 5.0 m; however, in the core channels, it ranges from 5.0 m to 11.5 m [8].

Sampling was carried out in three wetlands of Punjab, Pakistan, including two main Ramsar sites. Site I was Chashma barrage (District Mianwali: 32°25′4″ N; 71°26′19″ E), Site II was Taunsa barrage (District Muzaffargarh: 30°30′46″ N; 70°50′57″ E), and Site III was Trimmu barrage (District Jhang: 31°10′0″ N; 72°7′60″ E) (Figure 1).

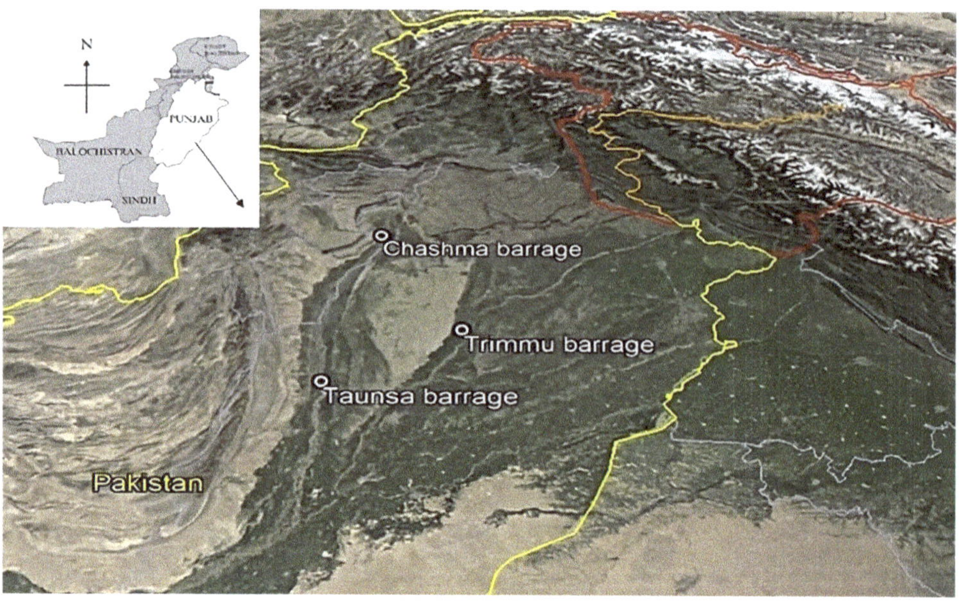

Figure 1. Location of sampling sites.

2.2. Collection and Preparation of Tissues

Heart and liver samples of *Anas crecca* (n = 25), *Fulica atra* (n = 26), and *Anas strepera* (n = 23) were obtained from licensed hunters between autumn 2021 and spring 2022. These birds were selected for this study because they serve as indicators of environmental health due to their wide distribution, ability to accumulate pollutants, ecological significance in wetland habitats, role as indicator species, and international significance for cross-border conservation efforts. Samples were packed in sterilized bags and put in a nitrogen environment for safe transportation. In the laboratory, the tissues were dissected with the help of an autoclaved stainless-steel knife, washed with double distilled water, and kept at −20 °C until digestion. On a thermolyte plate set at 200 °C, one gram of each tissue was digested until it was colorless using 10 mL of HNO_3 (65%) and 5 mL of $HClO_4$ (70%). Deionized water was added to the digested solution to dilute it to a level of 25 mL.

2.3. Quantification of Heavy Metals in Samples

A Thermo Fisher Scientific (Cambridge CB5 8BZ, UK) iCE 3000 Series flame atomic absorption spectrophotometer was used to measure the content of heavy metals. The Institute of Zoology, University of the Punjab, Lahore, Pakistan's Local Bioethical Committee accepted the rules that were followed for every experimental activity (Ref. D/1449/UZ). The standard reference material (SRM) used to check the accuracy of heavy metal analysis was NIST-SRM 1570 (National Institute of Standards and Technology, Gaithersburg, MD, USA). To determine the limit of detection (LOD) and limit of quantification (LOQ), 25 blanks were measured. LOQs and LODs were estimated by the method of [9]. The LODs for Cd, Co, Ni, Pb, Cu, Mn, and Zn were 0.001, 0.002, 0.001, 0.004, 0.003, 0.04, and 0.2 µg/g, respectively, while their LOQs were 0.003, 0.01, 0.003, 0.01, 0.01, 0.1, and 0.67, respectively. The minimum criterion for repeatability was 1%. When the values exceeded this limit, the analysis was repeated.

2.4. Statistical Analysis

Mean concentrations ± SD of metals for both organs were calculated by applying basic descriptive statistics on SPSS version 22. Differences in heavy metal burdens between sites, species, and tissues were assessed by using a one-way ANOVA test on Minitab 17.

2.5. Histopathological Analysis

Heart and liver tissues, removed from the sampled organisms, were stored in 10% formalin and stained with eosin and hematoxylin. The tissues were dehydrated by passing through an ethanol gradient (80%, 90%, and 100%), kept in cedar wood oil for clearing, placed in paralast, and kept in the incubator at 60 °C for 30 min. Then, the paraplast was replaced with a new one, and the samples were again kept in the incubator at 60 °C for 12 h. This procedure was repeated till the tissues became clear. For each tissue, a box block was made and then firmly fixed into a plastic caster. By using rotatory microtone, 2 to 3 µm thick segments of each tissue were cut down, moved to clear slides, stretched on Fischer slides, and warmed.

The tissues were deparaffinized by using xylene. Then, a 50–100% ethanol dilution was used for rehydration purposes. Tap water was used to wash the slides, and hematoxylin staining was performed. The slides were washed again to remove excess stains and then stained with eosin. After completely dehydrating the slides in absolute alcohol, two drops of DPX mountant were placed on each slide, and each slide was covered with a cover slip. The slides were dried and placed under a microscope attached to a digital camera at 40 X resolution.

3. Results

3.1. Comparison of Heavy Metals between Sites

The mean concentration of Cu, Pb, Ni, Cd, Zn, Co, and Mn in water at the three sampling sites and the WHO standards for these heavy metals are presented in Table 1. The concentrations of these heavy metals in the heart samples of *A. crecca*, *A. strepera*, and *F. atra* is presented in Table 2. In the heart samples, only the mean concentration of Ni ($p < 0.05$) was significantly higher in *F. atra* at Trimmu barrage. In *A. strepera*, the concentrations of Cu, Pb ($p < 0.05$), and Co ($p < 0.01$) were significantly higher at Chashma barrage, while the concentrations of Ni and Cd ($p < 0.05$) were significantly higher at Trimmu barrage. In *A. crecca*, the mean values of Cd ($p < 0.01$) and Mn ($p < 0.05$) were higher at Trimmu barrage, and that of Cu ($p < 0.05$) was higher at Taunsa barrage. In the liver samples, when mean concentrations were compared, Cu and Ni ($p < 0.05$) were significantly higher at Chashma barrage and Taunsa barrage, respectively, in *F. atra*. Similarly, in *A. strepera*, Ni ($p < 0.01$) and Zn ($p < 0.05$) were higher at Taunsa barrage and Chashma barrage, respectively. Lastly, Co ($p < 0.01$) was higher at Chashma barrage in *A. crecca* (Table 3).

Table 1. WHO criteria for heavy metals in Pakistan and the concentration of heavy metals in water samples from three sampling sites.

Sample	Site	Cu	Ni	Zn	Pb	Cd	Co	Mn
WHO standards for Pakistan 2004 (mg/L)		2	0.02	5	2.5	0.003	0.001	0.20
water (mg/L)	Taunsa barrage	0.26 ± 0.13	0.13 ± 0.039	0.297 ± 0.07	0.21 ± 0.17	0.011 ± 0.01	0.22 ± 0.34	0.053 ± 0.075
	Chashma barrage	0.019 ± 0.007	0.065 ± 0.058	0.316 ± 0.20	0.084 ± 0.04	0.003 ± 0.00	0.006 ± 0.007	0.02 ± 0.008
	Trimmu barrage	0.25 ± 0.07	0.285 ± 0.22	0.39 ± 0.019	0.69 ± 0.30	0.025 ± 0.018	0.004 ± 0.001	0.73 ± 0.16

Table 2. Mean ± S.D (μg/g) values of selected metals in hearts of water birds.

Metal	*F. atra*			*A. strepera*			*A. crecca*		
	Site I [a] N = 9	Site II [b] N = 6	Site III [c] N = 11	Site I N = 7	Site II N = 6	Site III N = 10	Site I N = 10	Site II N = 8	Site III N = 7
Cu	9.44 ± 5.16	10.14 ± 6.02	13.77 ± 2.42	144.6 ± 144.0	17.67 ± 4.13	11.43 ± 2.54	12.79 ± 3.09	54.3 ± 50.4	8.33 ± 4.33
		NS			$p < 0.05$			$p < 0.05$	
Ni	25.61 ± 15.79	12.75 ± 6.48	41.33 ± 18.20	14.59 ± 5.45	17.51 ± 14.35	32.67 ± 10.35	17.64 ± 17.73	15.16 ± 9.83	12.24 ± 11.39
		$p < 0.05$			$p < 0.05$			NS	
Zn	282.5 ± 150.8	304.2 ± 52.7	319.9 ± 215.1	310.4 ± 187.7	427.7 ± 190.1	298.5 ± 139.1	214.3 ± 132.8	217.8 ± 155.4	146.2 ± 104.7
		NS			NS			NS	
Pb	27.4 ± 30.7	24.48 ± 11.77	35.26 ± 13.56	54.76 ± 17.65	15.11 ± 8.58	43.2 ± 31.4	17.5 ± 10.77	23.63 ± 18.87	12.19 ± 13.52
		NS			$p < 0.05$			NS	
Cd	4.94 ± 2.51	14.40 ± 12.42	16.22 ± 8.72	7.62 ± 2.4	18.46 ± 13.30	24.18 ± 12.86	7.59 ± 1.02	6.03 ± 2.25	20.81 ± 11.59
		NS			$p < 0.05$			$p < 0.01$	
Co	1.14 ± 1.18	1.76 ± 1.48	0.76 ± 0.44	2.59 ± 0.53	2.18 ± 1.21	0.54 ± 0.14	1.41 ± 0.6	2.06 ± 1.96	1.31 ± 0.76
		NS			$p < 0.01$			NS	
Mn	115.31 ± 18.63	106.98 ± 20.65	94.97 ± 2.85	91.33 ± 6.29	95.25 ± 10.38	112 ± 30.8	117.08 ± 13.06	100.96 ± 15.43	124.33 ± 9.53
		NS			NS			$p < 0.05$	

Note: One-way ANOVA test was applied to compare mean values. [a] Site I: Chashma barrage; [b] Site II: Taunsa barrage; [c] Site III: Trimmu barrage; NS = non-significant.

Table 3. Mean ± S.D (μg/g) values of selected metals in livers of water birds.

Metal	*F. atra*			*A. strepera*			*A. crecca*		
	Site I [a] N = 9	**Site II** [b] N = 6	**Site III** [c] N = 11	**Site I** N = 7	**Site II** N = 6	**Site III** N = 10	**Site I** N = 10	**Site II** N = 8	**Site III** N = 7
Cu	61.4 ± 51.6	16.58 ± 9	8.18 ± 6.61	94 ± 84.2	114.7 ± 109.4	117.2 ± 116.6	22.46 ± 10.39	82.9 ± 113.7	27.82 ± 5.22
		$p < 0.05$			NS			NS	
Ni	27.47 ± 17.33	39.25 ± 20.22	13.22 ± 8.49	50.19 ± 6.24	67.8 ± 31	26.14 ± 3.88	32.83 ± 18.74	53.9 ± 33.1	78.5 ± 52
		$p < 0.05$			$p < 0.01$			NS	
Zn	503 ± 181.9	521 ± 144.6	363.3 ± 82.6	298.2 ± 205.3	166.5 ± 47.9	85.87 ± 10.83	269 ± 259	104.2 ± 41.4	116 ± 35.7
		NS			$p < 0.05$			NS	
Pb	24.95 ± 15.98	24.81 ± 20.84	28.47 ± 21.01	20.05 ± 6.18	9.85 ± 2.35	9.94 ± 0.57	13.19 ± 7.02	14.65 ± 7.22	12.49 ± 4.62
		NS			$p < 0.01$			NS	
Cd	14.14 ± 9.14	9.09 ± 0.88	19.2 ± 11.78	2.3 ± 2.94	0.39 ± 0.21	0.17 ± 0.04	4.17 ± 4.41	1.47 ± 3.19	1.35 ± 2.71
		NS			NS			NS	
Co	1.23 ± 0.83	1.62 ± 1.14	1.08 ± 1.04	2.81 ± 1.34	2.16 ± 1.18	2.66 ± 0.48	3.16 ± 1.28	1.02 ± 0.46	2.24 ± 0.95
		NS			NS			$p < 0.01$	
Mn	126.88 ± 12.28	115.73 ± 11.1	120.28 ± 11.73	179.7 ± 87.1	132 ± 47.5	66.15 ± 3.41	96.42 ± 15.62	79.61 ± 20.86	89.35 ± 15.68
		NS			$p < 0.05$			NS	

Note: One-way ANOVA test was applied to compare mean values. [a] Site I: Chashma barrage; [b] Site II: Taunsa barrage; [c] Site III: Trimmu barrage; NS = non-significant.

3.2. Comparison of Heavy Metals between Tissues

When the mean concentrations between the heart and liver samples were compared, the concentrations of metals were higher in the livers of water birds, but in *A. strepera*, they were higher in the heart tissues (Figures 2–4). Ni was significantly ($p < 0.05$) higher in the livers, except for *F. atra*, from Trimmu barrage, and the Cd concentration was significantly ($p < 0.05$) greater in the hearts when compared with the livers.

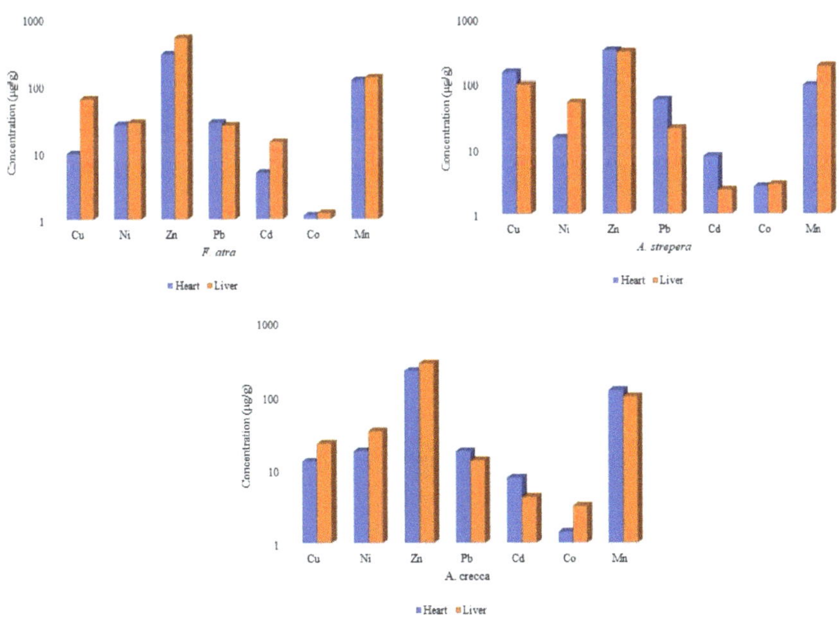

Figure 2. Heavy metals (μg/g) in hearts and livers of water birds from Chashma barrage.

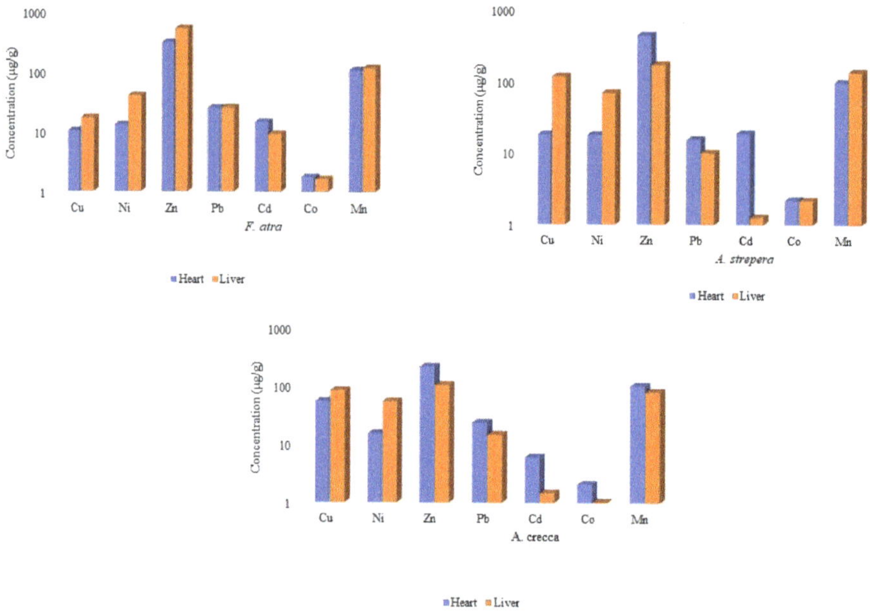

Figure 3. Heavy metals (µg/g) in hearts and livers of water birds from Taunsa barrage.

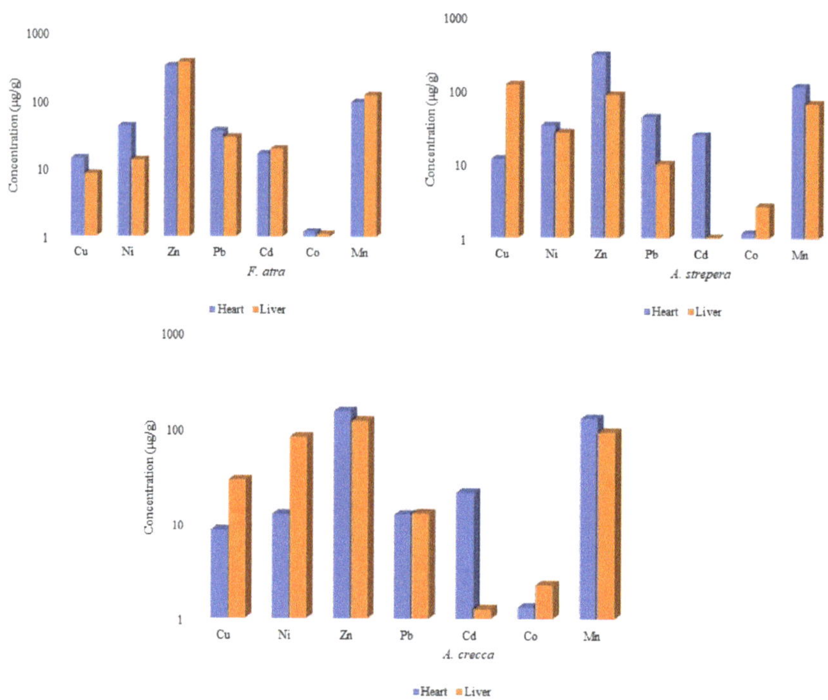

Figure 4. Heavy metals (µg/g) in hearts and livers of water birds from Trimmu barrage.

3.3. Comparison of Heavy Metals between Species

Among the species, *A. strepera* accumulated metals in the highest concentration ($p < 0.05$) at Chashma barrage and Taunsa barrage, while *F. atra* accumulated metals in the highest concentration at Trimmu barrage (Figures 5 and 6). *A. strepera* was the species in which Zn (427.7 µg/g), Cu (144.6 µg/g), Pb (54.76 µg/g), Cd (24.18 µg/g), and Co (2.59 µg/g) were found to be most abundant, while, *F. atra* showed remarkable differences in the concentrations of Zn, Cd, Pb, and Mn in the livers as compared to both other species.

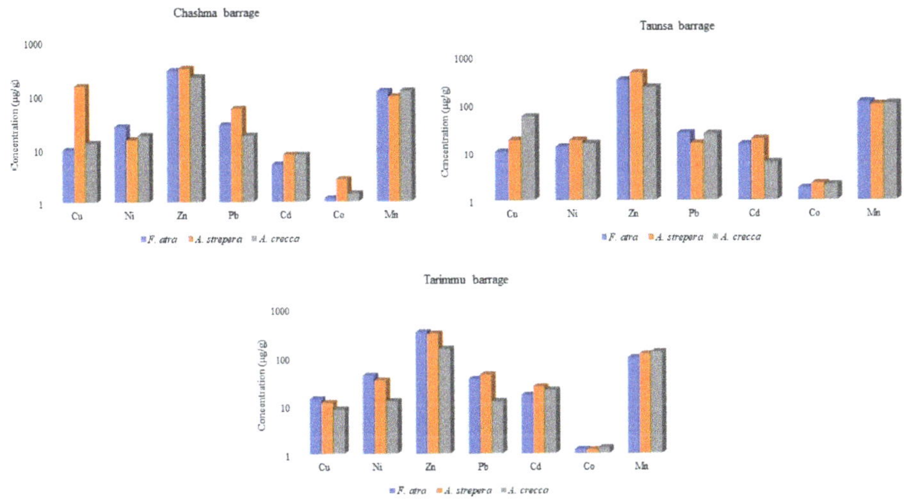

Figure 5. Comparison of heavy metals (µg/g) in hearts of water birds.

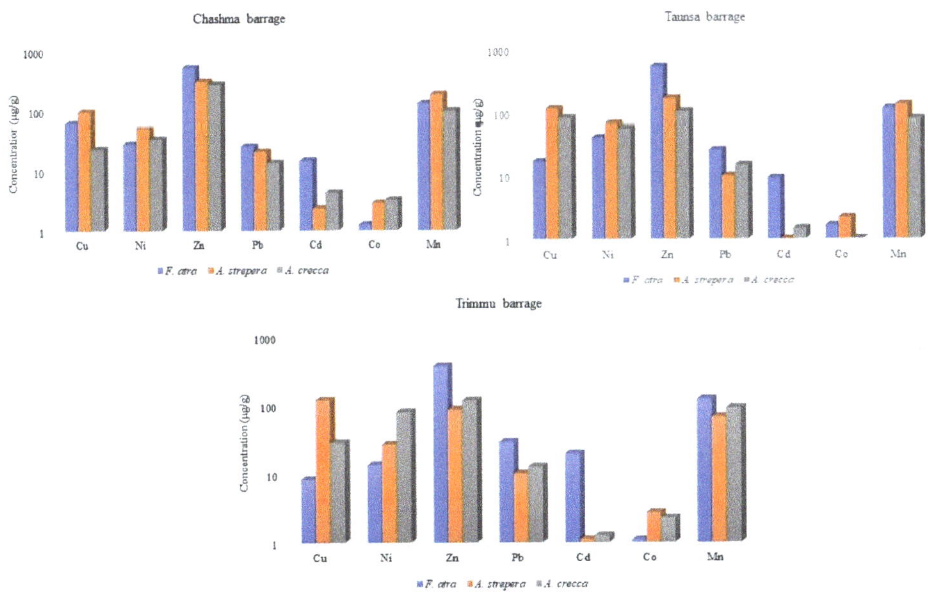

Figure 6. Comparison of heavy metals (µg/g) in livers of water birds.

3.4. Histopathological Analysis

3.4.1. Heart

Cardiomyocytes and striations in myocardial fibers were recorded for histological sections of *A. crecca* hearts, while myocarditis (arrow) was evident in the hearts of *A. crecca* from Chashma barrage (spring), Taunsa, and Trimmu barrage (autumn) (Figure 7).

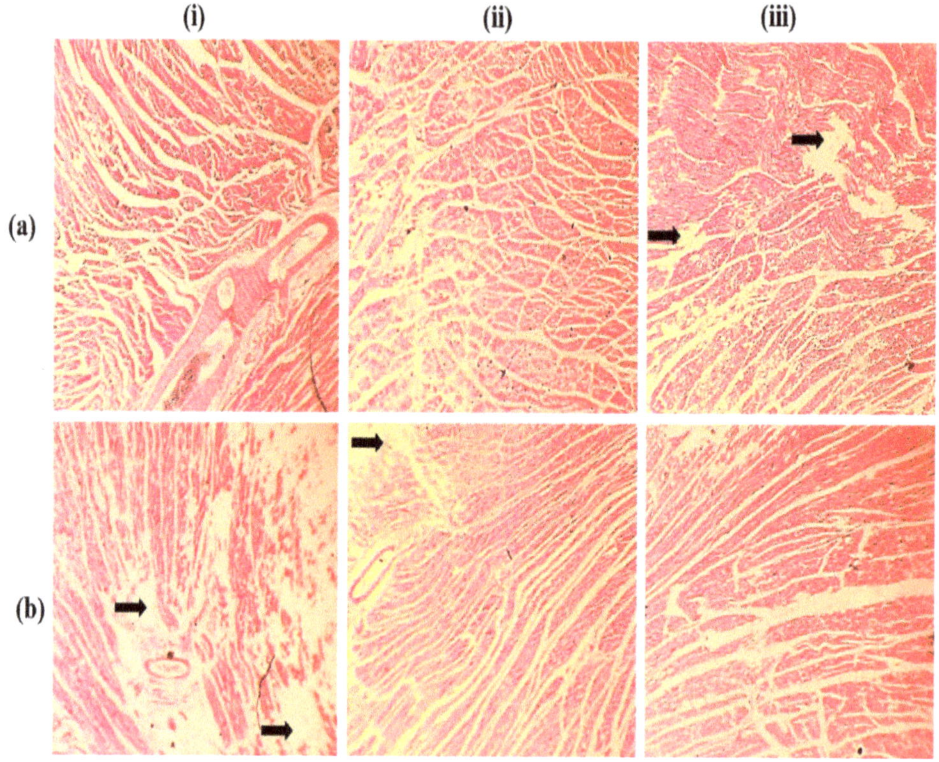

Figure 7. Photomicrographs of hearts of *A. crecca* from (**i**) Taunsa barrage, (**ii**) Trimmu barrage, and (**iii**) Chashma barrage during (**a**) autumn and (**b**) spring. H & E, magnification: 40× Myocarditis (arrow) was evident in the hearts of *A. crecca*.

Histological sections of hearts of *A. strepera* from all localities in both seasons depicted normal microarchitectures without any distortion of cardiomyocytes and striated muscle fibers, except for the degeneration of muscle fibers in a heart of *A. strepera* sample collected from Trimmu barrage in autumn. A widened interstitium and necrosis of myocardial fibers were also observed in the heart section of the same sample (Figure 8).

The histological sections of *F. atra* hearts from all localities in both seasons were found to be anomalous. Extensive myocarditis was observed in samples collected from Taunsa barrage (autumn), Trimmu barrage (in both autumn and spring), and Chashma barrage (spring), while a significantly widened interstitial space was observed in histological sections of *F. atra* hearts sampled from Chashma barrage (autumn) and Taunsa barrage (spring) (Figure 9).

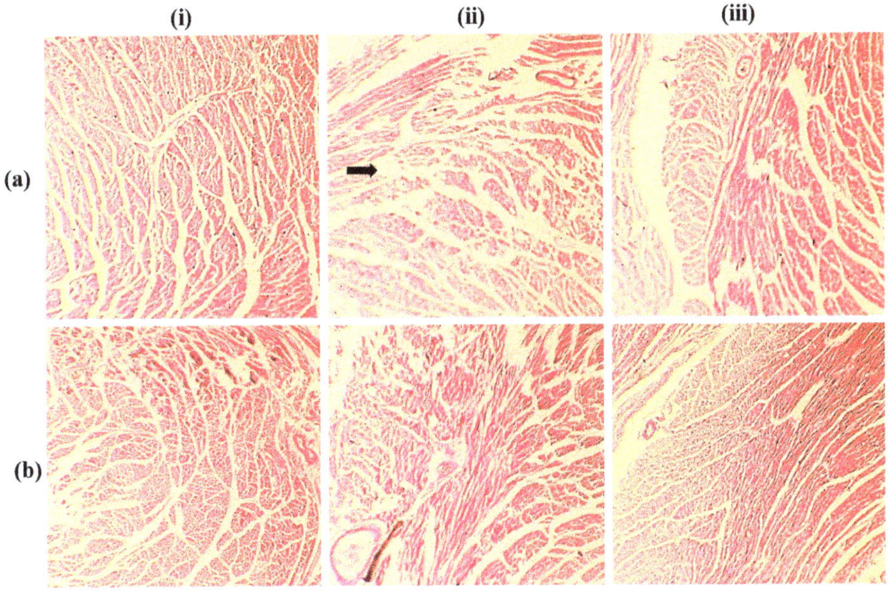

Figure 8. Photomicrographs of hearts of *A. strepera* from (**i**) Taunsa barrage, (**ii**) Trimmu barrage, and (**iii**) Chashma barrage during (**a**) autumn and (**b**) spring. H & E, magnification: 40×.

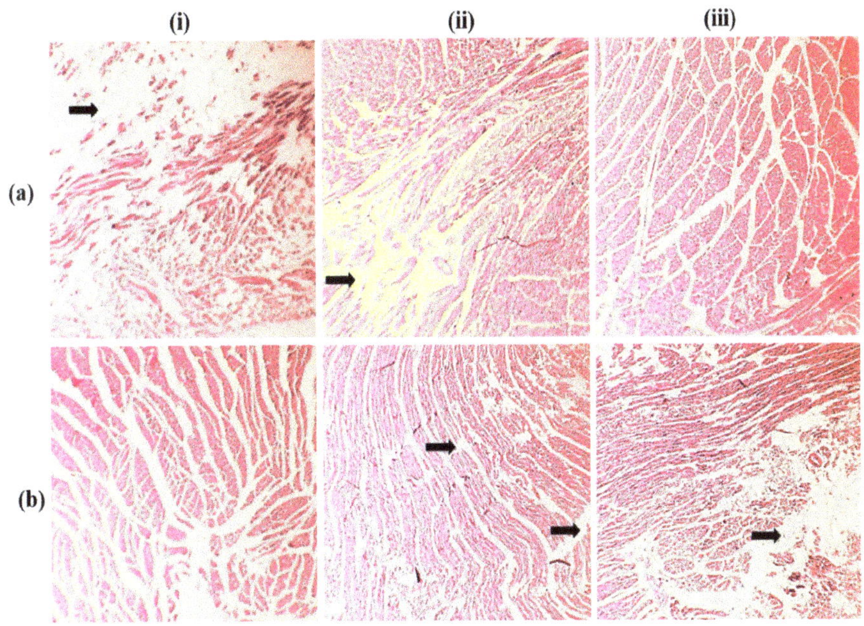

Figure 9. Photomicrographs of hearts of *F. atra* from (**i**) Taunsa barrage, (**ii**) Trimmu barrage, and (**iii**) Chashma barrage during (**a**) autumn and (**b**) spring. H & E, magnification: 40× (black arrow showed myocarditis).

3.4.2. Liver

Melan macrophage aggregations (*), the dilation of diss space (small arrow), and degenerative hepatocytes (larger black arrow) appeared in histological sections of *A. crecca* livers sampled from Chashma barrage (spring) and Taunsa and Trimmu barrage (autumn) (Figure 10).

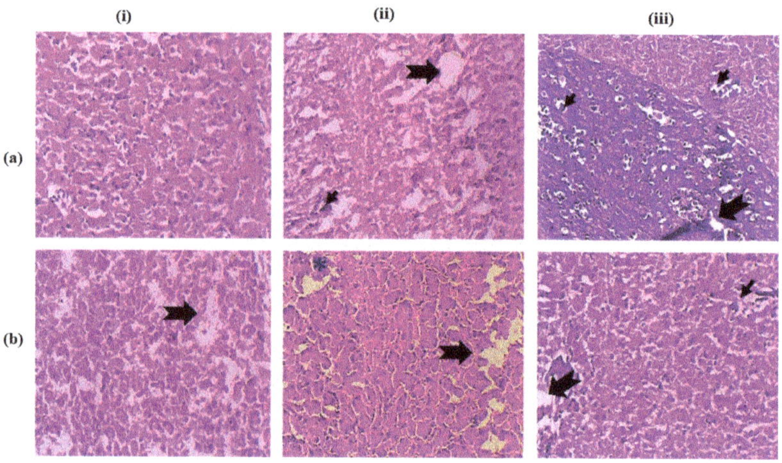

Figure 10. Photomicrographs of livers of *A. crecca* from (**i**) Taunsa barrage, (**ii**) Trimmu barrage, and (**iii**) Chashma barrage during (**a**) autumn and (**b**) spring. H & E, magnification: 40× (Melan macrophage aggregations (*), the dilation of diss space (small arrow), and degenerative hepatocytes (larger black arrow).

Necrosis (N) and vacuolation (V) appeared in histological sections of *A. strepera* livers sampled from Chashma barrage (spring) and Taunsa and Trimmu barrage (autumn) (Figure 11).

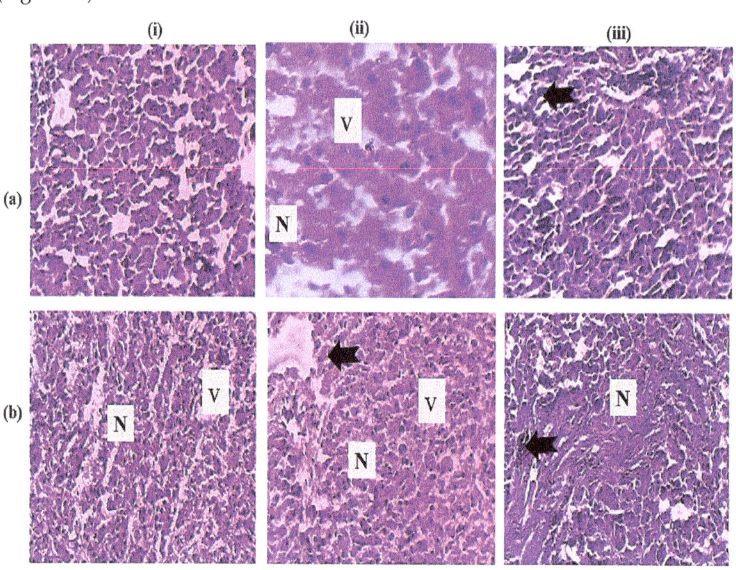

Figure 11. Photomicrographs of livers of *A. strepera* from (**i**) Taunsa barrage, (**ii**) Trimmu barrage, and (**iii**) Chashma barrage during (**a**) autumn and (**b**) spring. H & E, magnification: 40× (V = vacuolation and N = Necrosis).

Necrosis (N), vacuolation (V), and pyclotic nuclei necrosis (PN) appeared in histological sections of *F. atra* livers sampled from Chashma barrage (spring) and Taunsa and Trimmu barrage (autumn) (Figure 12).

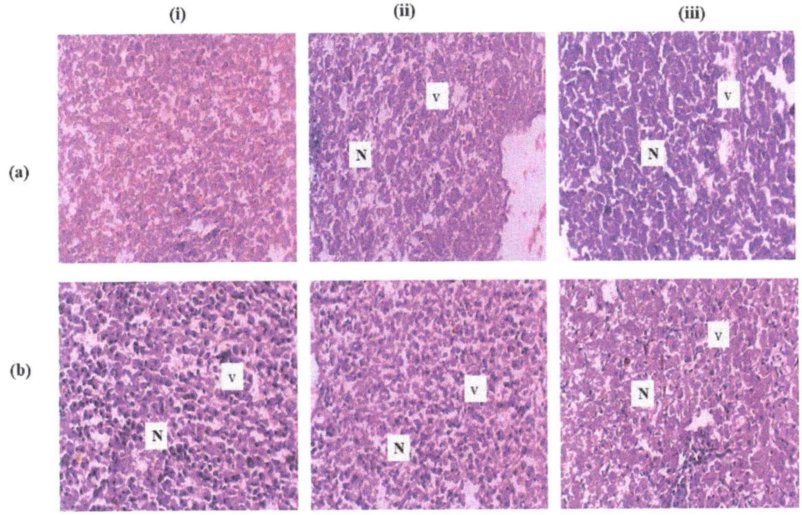

Figure 12. Photomicrographs of livers of *F. atra* from (**i**) Taunsa barrage, (**ii**) Trimmu barrage, and (**iii**) Chashma barrage during (**a**) autumn and (**b**) spring. H & E, magnification: 40× (V = vacuolation and N = Necrosis).

4. Discussion

In the present study, the concentrations of metals varied with the sampling sites, species, and tissues. It was found that Cd had the lowest concentration among the metals (0.17 μg/g in the livers of *A. strepera* at Trimmu barrage), and Zn had the highest (521 μg/g in the livers of *F. atra* at Taunsa barrage).

4.1. Metal Concentrations in Tissues

In the heart samples, Cu ranged between 8.33 μg/g and 144.6 μg/g, comparable to the findings of Ruelas-Inzunza and Páez-Osuna [10] (8.6 μg/g–41 μg/g) and Akoto et al. [11] (16 μg/g–21.2 μg/g). The average amount of Cu in the liver lies within the range of 8.18 μg/g and 117.2 μg/g. The liver detoxifies pollutants and other harmful compounds that enter the body, making it more vulnerable to the toxic effects and buildup of heavy metals in tissues [12]. For water birds, Cu in concentrations ranging between 270 mg/kg and 1300 mg/kg dry weight is toxic, with an acute toxicity value of 700 mg/kg [13]. It was found that the concentration of Cu was lower in all samples than the acute toxicity limits but higher than the normal physiologically required concentration. In the present study, the concentration of Zn was lower than its toxic limits (>440 mg/kg dry weight) [14], except in the livers of *F. atra* (521 μg/g). As Zn is actively regulated in birds because it comes through diet, it did not affect the livers. But when it reaches higher limits than the metabolic needs of the body, birds are threatened by its health-damaging effects [15]. According to our findings, Ni was comparatively higher in the livers in comparison to the hearts. The average Ni concentration ranged from 13.22 μg/g to 78.5 μg/g, higher than the values reported by the authors of [16].

The mean Cd concentration varied from 4.94 μg/g to 24.18 μg/g and 0.17 μg/g to 19.2 μg/g for the hearts and livers, respectively. These concentrations were higher than those reported in *F. americana* (3.3 μg/g) [10], *A. crecca* (0.55 μg/g) [17], *F. atra* (0.85 μg/g) [18], *A. crecca* (0.83 μg/g), and *A. strepera* (1.08 μg/g) [19]. Water birds are more at risk from the toxicity of Pb because they either swallow lead shots or they feed upon cereals and weeds grown in fields irrigated by Pb-contaminated waters [20]. In the present study, the Pb concentration was highest in the hearts of *A. strepera* (54.76 μg/g) and livers of *F. atra* (28.47 μg/g). These concentrations were far greater than the reported threshold limits of poisoning (<7.5 μg/g wet weight) [21]. The concentration of Co ranged from 1.02 μg/g to 3.16 μg/g and 0.54 μg/g to 2.59 μg/g in the livers and hearts, respectively, in the present investigation. These values are greater than those reported by Yohannes et al. [22] (i.e., 0.003 μg/g for hearts and 0.11–1.4 μg/g for livers). The main anthropogenic sources of enhanced Co in natural wetlands are domestic, industrial, and agricultural wastewater discharge [23]. The dietary route is the main route that facilitates the accumulation of Mn in the bodies of water birds. The Mn concentration reported in the present study was found to be far greater when compared with the findings of Ruelas-Inzunza and Páez-Osuna [10], Kim et al. [24], Kim and Oh [25], and Sujak et al. [26]. Long-term exposure to heavy metals may cause disruptive behavior and reduced resistance against diseases [27].

4.2. Comparison between Sites

A general trend of increasing heavy metals concentrations at Chashma barrage > Taunsa barrage > Trimmu barrage in the heart samples and Taunsa barrage > Chashma barrage > Trimmu barrage in the liver samples was found. The mean metal concentrations were high at Trimmu barrage perhaps because Trimmu barrage receives wastewater containing pollutants (industrial, domestic, agricultural) from local and distant sources. The possible source of the pollutants in Trimmu barrage is wastewater discharged from the cities of Gujrat, Sialkot, and Faisalabad into the Chenab River, in these cities, major industries (such as the fertilizer industry, paper and pulp products industry, leather tanneries, and textile industries) are prevalent [23].

Chashma barrage and Taunsa barrage are situated on the Indus River and play an important role in irrigation and fishing, but municipal discharge, agricultural runoff, industrial effluents, and sewage contaminate their water, which generates public health issues. The untreated industrial wastewater discharge magnifies the level of contamination of the Indus River's water. The human population that lives by the river increases its pollution by dumping its excretory and domestic waste in the river. A considerable concentration of heavy metals is present in these effluents, which threaten the health of the inhabitants of aquatic bodies [28].

4.3. Comparison between Species

Among the species, the mean Cu was highest in *A. strepera*, which is attributed to its larger body size. The increasing order of Ni was as follows: *F. atra* < *A. strepera* < *A. crecca*. The mean Ni concentrations were higher than those reported by Lebedeva [29] and Van Eeden [30]. The mean concentration of Zn ranged between 85.87 μg/g and 521 μg/g, higher than the values in various studies [19,31,32]. The mean lead concentration was highest in *A. strepera* heart samples, while in *F. atra*, it was higher in the livers. *F. atra* habitually ingest more Pb shots [23], which might be the cause of the high Pb levels in this species, while *A. strepera* and *A. crecca* inhabit the same areas, but the body size difference affects their Pb accumulation. Cadmium concentrations below 3 μg/g dry weight are considered to be safe for birds, but we found higher levels than these in all samples except in the livers of *A. strepera* from Taunsa barrage. The mean concentration of Cd in the present study is far higher than that reported in *F. atra*, *A. crecca*, and *A. strepera* (0.002–0.026 μg/g) by Mansouri and Majnoni [18], and Sinkakarimi, Binkowski, Hassanpour, Rajaei, Ahmadpour and Levengood [19], respectively. Discharges from industrial and municipal sources are the main cause of cobalt in wetlands [23]. The increasing order of mean Co was found to be *F. atra* < *A. crecca* < *A. strepera*. The mean Mn concentration ranged from 66.15 μg/g to 179.7 μg/g, which is higher than that reported by Aloupi et al. [33], Mateo and Guitart [32], Lebedeva [29], and the authors of [23].

4.4. Histopathology

Cobbina et al. [34] found that co-exposure to many metals resulted in a higher level of toxicity when compared to exposure to each metal individually. In the current study, cardiomyocytes and striations in myocardial fibers appeared in histological sections of the heart samples, and necrosis (N), vacuolation (V), and pyclotic nuclei necrosis (PN) appeared in histological sections of *A. strepera* livers sampled from Chashma barrage (spring) and Taunsa and Trimmu barrage (autumn). No similar studies have been conducted previously; however, heavy metal exposure has been shown to cause abnormally organized myofibrils, apoptosis, and vacuolization in the heart tissue of rats [35,36]. According to the results of Jayawardena et al. [37], heavy metals caused alternations in the liver histology of amphibians, and Riaz, Nisa, Anjum, Butt, Mehmood, Riaz and Akhtar [35] reported necrosis and cell degeneration in liver tissues due to co-exposure to heavy metals.

5. Conclusions

The wetlands of Punjab are under a heavy burden of heavy metal pollution. The concentration of heavy metals in aquatic bodies reported in the present study is within the toxic range for water birds. Furthermore, heavy metals caused health issues in the sampled birds, as alternations in cardiac tissues and hepatic tissues were observed in photomicrographs of both tissue types after they were stained with E & H. The damaged tissues are an indication of the severity of toxicity faced by these wintering species, which could reduce their reproductive potential and challenge their survival. In addition, the polluted wetlands could induce behavioral changes, including to their migration patterns.

Author Contributions: Conceptualization, M.S.K. and M.A.A.; methodology, M.W.; software, S.S.; validation, S.-M.K. and A.A.A.-G. formal analysis, A.U.; investigation, M.A.A.; resources, S.-M.K.; data curation, A.A.A.-G.; writing—original draft preparation, S.S. and A.U.; writing—review and editing, M.S.K., M.W. and M.A.A.; visualization, S.-M.K.; supervision, M.S.K. and I.-J.L.; project administration, A.A.A.-G. and I.-J.L.; funding acquisition, S.S and I.-J.L. All authors have read and agreed to the published version of the manuscript.

Funding: This research was supported by the Basic Science Research Program through the National Research Foundation of Korea (NRF), funded by the Ministry of Education (RS-2023-00243438).

Data Availability Statement: Data are contained within the article.

Acknowledgments: The authors extend their appreciation to the Researchers Supporting Project number (RSP2024R483), King Saud University, Riyadh, Saudi Arabia.

Conflicts of Interest: The author declares no conflicts of interest.

References

1. Tylmann, W.; Gołębiewski, R.; Woźniak, P.P.; Czarnecka, K. Heavy metals in sediments as evidence for recent pollution and quasi-estuarine processes: An example from Lake Druzno, Poland. *Environ. Geol.* **2007**, *53*, 35–46. [CrossRef]
2. Burger, J.; Diaz-Barriga, F.; Marafante, E.; Pounds, J.; Robson, M. Methodologies to examine the importance of host factors in bioavailability of metals. *Ecotoxicol. Environ. Saf.* **2003**, *56*, 20–31. [CrossRef]
3. Peakall, D.; Burger, J. Methodologies for assessing exposure to metals: Speciation, bioavailability of metals, and ecological host factors. *Ecotoxicol. Environ. Saf.* **2003**, *56*, 110–121. [CrossRef] [PubMed]
4. Spiegel, H. Trace element accumulation in selected bioindicators exposed to emissions along the industrial facilities of Danube Lowland. *Turk. J. Chem.* **2002**, *26*, 815–824.
5. Mahboob, S.; al-Balawi, H.A.; Al-Misned, F.; Ahmad, Z.; Sultana, S. Study on Avian Diversity of Thal Desert (District Jhang), Punjab, Pakistan. *Life Sci.* **2013**, *10*, 1–8.
6. Mahboob, S. Diversity of avifauna of Trimmu barrage, district Jhang, Punjab, Pakistan. *Pak. J. Zool.* **2009**, *41*, 43–49.
7. Zaman, A.; Rafique, A.; Ashraf, A.; Mahmood, M.S.; Nahid, N.; Jabeen, F.; Sultana, S.; Sultana, T.; Ali, S.; Zihad, S.M.; et al. Diversity, Abundance, and Distribution of Avifauna in District Jhang, Pakistan. *J. Zool. Syst. Evol. Res.* **2022**, *2023*, 3308651. [CrossRef]
8. Mahmood-ul-Hassan, M. Population trend of the black coot (*Fulica atra*) in the Punjab, Pakistan during 1989 through 2008. *Pak. J. Zool.* **2011**, *43*, 665–671.
9. Eka, N.; Retno, S.; Rohman, A. Validation and quantitative analysis of cadmium and lead in snake fruit by flame atomic absorption spectrophotometry. *Int. Food Res. J.* **2012**, *19*, 937.

10. Ruelas-Inzunza, J.; Páez-Osuna, F. Trace metals in tissues of resident and migratory birds from a lagoon associated with an agricultural drainage basin (SE Gulf of California). *Arch. Environ. Contam. Toxicol.* **2004**, *47*, 117–125. [CrossRef]
11. Akoto, O.; Bortey-Sam, N.; Ikenaka, Y.; Nakayama, S.M.; Baidoo, E.; Yohannes, Y.B.; Ishizuka, M. Contamination levels and sources of heavy metals and a metalloid in surface soils in the Kumasi Metropolis, Ghana. *J. Health Pollut.* **2017**, *7*, 28–39. [CrossRef] [PubMed]
12. Lucia, M.; André, J.-M.; Bernadet, M.-D.; Gontier, K.; Gérard, G.; Davail, S. Concentrations of metals (zinc, copper, cadmium, and mercury) in three domestic ducks in France: Pekin, muscovy, and mule ducks. *J. Agric. Food Chem.* **2008**, *56*, 281–288. [CrossRef] [PubMed]
13. Isanhart, J.P.; Wu, H.; Pandher, K.; MacRae, R.K.; Cox, S.B.; Hooper, M.J. Behavioral, clinical, and pathological characterization of acid metalliferous water toxicity in mallards. *Arch. Environ. Contam. Toxicol.* **2011**, *61*, 653–667. [CrossRef] [PubMed]
14. Beyer, W.N.; Dalgarn, J.; Dudding, S.; French, J.B.; Mateo, R.; Miesner, J.; Sileo, L.; Spann, J. Zinc and lead poisoning in wild birds in the Tri-State Mining District (Oklahoma, Kansas, and Missouri). *Arch. Environ. Contam. Toxicol.* **2004**, *48*, 108–117. [CrossRef] [PubMed]
15. Sileo, L.; Nelson Beyer, W.; Mateo, R. Pancreatitis in wild zinc-poisoned waterfowl. *Avian Pathol.* **2003**, *32*, 655–660. [CrossRef] [PubMed]
16. Deng, H.; Zhang, Z.; Chang, C.; Wang, Y. Trace metal concentration in great tit (*Parus major*) and greenfinch (*Carduelis sinica*) at the Western Mountains of Beijing, China. *Environ. Pollut.* **2007**, *148*, 620–626. [CrossRef] [PubMed]
17. Fedynich, A.M.; Ballard, B.M.; McBride, T.J.; Estrella, J.A.; Garvon, J.M.; Hooper, M.J. Arsenic, cadmium, copper, lead, and selenium in migrating blue-winged teal (*Anas discors* L.). *Arch. Environ. Contam. Toxicol.* **2007**, *53*, 662–666. [CrossRef] [PubMed]
18. Mansouri, B.; Majnoni, F. Comparison of the metal concentrations in organs of two bird species from western of Iran. *Bull. Environ. Contam.* **2014**, *92*, 433–439. [CrossRef] [PubMed]
19. Sinkakarimi, M.-H.; Binkowski, L.J.; Hassanpour, M.; Rajaei, G.; Ahmadpour, M.; Levengood, J.M. Metal concentrations in tissues of gadwall and common teal from Miankaleh and Gomishan International wetlands, Iran. *Biol. Trace Elem. Res.* **2018**, *185*, 177–184. [CrossRef]
20. Binkowski, Ł.J. The influence of environmental conditions on lead transfer from spent gunshot to sediments and water: Other routes for Pb poisoning. *Chemosphere* **2017**, *187*, 330–337. [CrossRef]
21. Pain, D. Lead in waterfowl. In *Environmental Contaminants in Wildlife: Interpreting Tissue Concentrations*; Beyer, W.M., Heinz, G.H., Redman-Norwood, A.W., Eds.; Lewis Publishers: Boca Raton, FL, USA, 1996.
22. Yohannes, Y.B.; Ikenaka, Y.; Nakayama, S.M.M.; Mizukawa, H.; Ishizuka, M. Trace element contamination in tissues of four bird species from the Rift Valley Region, Ethiopia. *Bull. Environ. Contam. Toxicol.* **2017**, *98*, 172–177. [CrossRef] [PubMed]
23. Ullah, K.; Hashmi, M.Z.; Malik, R.N. Heavy-metal levels in feathers of cattle egret and their surrounding environment: A case of the Punjab Province, Pakistan. *Arch. Environ. Contam. Toxicol.* **2014**, *66*, 139–153. [CrossRef] [PubMed]
24. Kim, J.; Koo, T.-H.; Oh, J.-M. Monitoring of heavy metal contamination using tissues of two Ardeids chicks, Korea. *Bull. Environ. Contam. Toxicol.* **2010**, *84*, 754–758. [CrossRef] [PubMed]
25. Kim, J.; Oh, J.-M. Metal levels in livers of waterfowl from Korea. *Ecotoxicol. Environ. Saf.* **2012**, *78*, 162–169. [CrossRef] [PubMed]
26. Sujak, A.; Wiącek, D.; Jakubas, D.; Komosa, A.; Kitowski, I. Mallards Anas platyrhynchos shot in Eastern Poland: Ecological risk evaluated by analysis of trace elements in liver. *Hum. Ecol. Risk Assess. Int. J.* **2019**, *25*, 2116–2132. [CrossRef]
27. Dauwe, T.; Janssens, E.; Pinxten, R.; Eens, M. The reproductive success and quality of blue tits (*Parus caeruleus*) in a heavy metal pollution gradient. *Environ. Pollut.* **2005**, *136*, 243–251. [CrossRef]
28. Jabeen, F.; Chaudhry, A.S. Environmental impacts of anthropogenic activities on the mineral uptake in Oreochromis mossambicus from Indus River in Pakistan. *Environ. Monit. Assess.* **2010**, *166*, 641–651. [CrossRef] [PubMed]
29. Lebedeva, N. Accumulation of heavy metals by birds in the southwest of Russia. *Russ. J. Ecol.* **1997**, *28*, 41–46.
30. Van Eeden, P.H. Metal concentrations in selected organs and tissues of five Red-knobbed Coot (*Fulica cristata*) populations. *Water SA* **2003**, *29*, 313–322. [CrossRef]
31. Carpene, E.; Serra, R.; Isani, G. Heavy metals in some species of waterfowl of northern Italy. *J. Wildl. Dis.* **1995**, *31*, 49–56. [CrossRef]
32. Mateo, R.; Guitart, R. Heavy metals in livers of waterbirds from Spain. *Arch. Environ. Contam. Toxicol.* **2003**, *44*, 0398–0404. [CrossRef] [PubMed]
33. Aloupi, M.; Karagianni, A.; Kazantzidis, S.; Akriotis, T. Heavy metals in liver and brain of waterfowl from the Evros Delta, Greece. *Arch. Environ. Contam. Toxicol.* **2017**, *72*, 215–234. [CrossRef] [PubMed]
34. Cobbina, S.J.; Chen, Y.; Zhou, Z.; Wu, X.; Zhao, T.; Zhang, Z.; Feng, W.; Wang, W.; Li, Q.; Wu, X.; et al. Toxicity assessment due to sub-chronic exposure to individual and mixtures of four toxic heavy metals. *J. Hazard. Mater.* **2015**, *294*, 109–120. [CrossRef] [PubMed]
35. Riaz, M.A.; Nisa, Z.U.; Anjum, M.S.; Butt, H.; Mehmood, A.; Riaz, A.; Akhtar, A.B.T. Assessment of metals induced histopathological and gene expression changes in different organs of non-diabetic and diabetic rats. *Sci. Rep.* **2020**, *10*, 5897. [CrossRef] [PubMed]

36. Xu, L.-H.; Mu, F.-F.; Zhao, J.-H.; He, Q.; Cao, C.-L.; Yang, H.; Liu, Q.; Liu, X.-H.; Sun, S.-J. Lead induces apoptosis and histone hyperacetylation in rat cardiovascular tissues. *PLoS ONE* **2015**, *10*, e0129091. [CrossRef]
37. Jayawardena, U.A.; Angunawela, P.; Wickramasinghe, D.D.; Ratnasooriya, W.D.; Udagama, P.V. Heavy metal–induced toxicity in the Indian green frog: Biochemical and histopathological alterations. *Environ. Toxicol. Chem.* **2017**, *36*, 2855–2867. [CrossRef]

water

MDPI

Article

Evaluating the Ecological Impact of Wastewater Discharges on Microbial and Contaminant Dynamics in Rivers

Wenjie Jing [1], Shahdev Sajnani [2], Mengting Zhou [1], Hongfei Zhu [1,*] and Ya Xu [3,*]

[1] College of Environmental Science and Engineering, Liaoning Technical University, Fuxin 123000, China; jingwenjie1998@163.com (W.J.); zhoumengting13@163.com (M.Z.)

[2] School of Environmental Science and Engineering, Tianjin University, Tianjin 300072, China; shahdevsajnani@tju.edu.cn

[3] State Environmental Protection Key Laboratory of Hazardous Waste Identification and Risk Control, Chinese Research Academy of Environmental Sciences, Beijing 100012, China

* Correspondence: zhuhongfei2008@163.com (H.Z.); xuya@craes.org.cn (Y.X.)

Abstract: This study focused on assessing the impact of pollutants discharged from the Fuxin Meng Wastewater Treatment Plant (FMWWTP) on the microbial communities in the Xi River; specific comparison between upstream and downstream regions was made. Water samples were obtained by mixing morning, midday, and evening sampling at two points: one upstream and one downstream, each 500 m from the FMWWTP outfall. Utilizing metagenomic sequencing in conjunction with measured conventional physical and chemical properties, the research aimed to elucidate differences in microbial community structure, metabolic functions, potential biological contamination, and antibiotic resistance gene prevalence. The findings indicated a notable decrease in microbial diversity downstream compared to upstream; this was influenced primarily by the effluent from FMWWTP. This disparity in microbial diversity was evident at various taxonomic levels, with downstream samples showing higher diversity at the phylum level than at the genus level. Furthermore, downstream microbial populations demonstrated a broader range of metabolic and functional genetic diversity. Interestingly, the abundance of metabolic systems was generally greater downstream, with the notable exception of energy metabolism. This could be attributed to the stress imposed on downstream microorganisms by organic chemicals discharged from the treatment plant, and this prompts an enhanced metabolic decomposition function. The study also uncovered significant levels of potential biological contamination and antibiotic resistance gene pollution. This was more pronounced downstream of FMWWTP. In conclusion, discharge from FMWWTP has a substantial impact on the microbial communities of the Xi River, and this underscores the urgent need to optimize wastewater treatment processes to better comply with environmental quality standards.

Keywords: wastewater; water treatment plant impact; metagenomic contamination analysis; microbial community dynamics; antibiotic resistance genes

1. Introduction

Wastewater Treatment Plants (WWTPs) play a crucial role in processing urban domestic wastewater. They do this by effectively reducing harmful substances to minimize environmental impact. Despite their importance, effluents from WWTPs often fail to meet set standards [1,2]. Historically, the pollution of rivers by WWTP effluents may have been underestimated due to belief in the self-purifying capacity of rivers. However, as wastewater discharge and industrialization has increased, rivers have been put under growing stress, which diminishes their self-purification ability. The impact of WWTP discharge exceeds the inherent capacity of rivers, and this directly affects vital water resources and the ecological balance.

Studies have identified WWTP effluents as major sources of nitrogen and phosphorus in rivers. There is a positive correlation between the concentration of nitrogen [3], phosphorus, total nitrogen (TN), and chemical oxygen demand (COD) [4] in rivers, and the volume of wastewater discharged. This effluent adversely affects the aquatic ecosystem [5], altering microbial community structures and promoting the proliferation of difficult-to-remove resistance genes [6]. Consequently, recipient rivers may suffer from biological and chemical pollution, spread of resistance genes, and bioaccumulation of trace pollutants [7,8]. This significantly harms water quality and affects healthy ecosystems. A comprehensive analysis is needed to understand the extent of the damage to water quality and ecology in these rivers.

This study uses both chemical and biological indicators to comprehensively analyze the impact of FMWWTP discharge on the Xi River. This method differs from traditional single-factor index analyses. Traditional culturing techniques cannot fully identify the complex microbial communities [9], therefore, potential microbial mechanisms affected by the effluent are often overlooked [10]. Metagenomics was, therefore, chosen to analyze the comprehensive impact of FMWWTP on the Xi River. Metagenomics, which is a reliable technique for analyzing the DNA of all environmental microbes [11], helps in the study of the composition, function, diversity, interactions, and genetic aspects of microbial communities in various environmental processes [12]. This approach avoids PCR biases, and offers advantages over 16S amplicon sequencing [13]. This integrated analysis, which combines metagenomic technology with physicochemical property indicators, provides a more accurate reflection of the pollution in the river from the FMWWTP; in comparison to traditional single-factor indicators.

This research primarily focuses on water quality changes resulting from river discharge, which emphasizes the impact on microbial communities. Through gene abundance analysis, we aim to unravel the underlying mechanisms driving community succession, and to identify traits and responses from a vast gene pool. This study uniquely considers the unilateral effects of WWTP effluent, as no other effluent sources were identified between our sampling sites. By examining changes in water quality, microbial community structure, metabolic functions, and anti-pollution genes, we endeavor to present a comprehensive overview of the effects of sewage discharge on the Xi River.

2. Materials and Methods

2.1. Sample Collection

Water samples were collected from the Xi River, situated near Fuxin in the Liaoning Province of China. The exact sampling coordinates are 42.052334° N, 121.747798° E, as depicted in Figure 1. Two sampling sites were chosen along the river: one 500 m upstream (XDU) and another 500 m downstream (XDD) from the Fuxin Meng wastewater treatment plant's outflow into the Xi River. This sampling approach was designed to investigate the impact of the wastewater treatment plant on the quality of river water in these areas.

Samples were collected three times: once in the morning, once at midday, and one in the evening. Each sample amounted to 1 L. After collection, they were immediately placed in ice bags and stored at 4 °C until the next day. Water quality was tested according to GB3838-2002 standards [14]. Temperature and pH were measured at the sampling sites. BOD5, COD, TP, TN and ammonia nitrogen (AN) were analyzed in the laboratory immediately after collection, as per GB3838-2002 methods. The results are presented in Table 1 [15].

Table 1. Physical and chemical properties of Xi River samples.

Physical and Chemical Properties	Temperature (°C)	pH	BOD_5 (mg/L)	COD (mg/L)	TP (mg/L)	TN (mg/L)	AN (mg/L)
Upstream	11.4	6.8	10	12.5	0.079	24.93	3.669
Downstream	10.9	6.9	43	100	0.242	25.13	0.524

Figure 1. Geographical sampling locations along the Xi River.

Studies have indicated that eutrophication is evident when phosphorus and nitrogen concentrations exceed 0.02 mg/L [16]. This suggests that the Xi River is in a eutrophic state. According to previous investigations, the composite water pollution index in the Xi River Basin has fluctuated between 0.2 and 0.4 over the past two decades. This indicates that there is a consistent pattern [17]. This index reflects the overall pollution level, including nutrients such as phosphorus and nitrogen, which contribute to eutrophication. Therefore, the data presented in this study reflect these broader environmental conditions.

2.2. DNA Extraction

Water samples were initially filtered through a 0.22 μm pore size membrane (polycarbonate membrane) to capture precipitate on the membrane surface. This precipitate was then used for DNA extraction, while the supernatant was separated by centrifugation and stored at −4 °C. The DNA extraction was performed using a Tiangen Biotech Co., Ltd. (Beijing, China) DNA kit. This yielded 30 μL of microbial DNA from each sample. The DNA samples were tested for quality using a NanoPhotometer (Implen, Munich, Germany), to ensure the DNA met stringent quality standards. High-quality DNA samples were subsequently packaged in ice bags and sent to Shanghai Major Biomedical Technology Co (Shanghai, China). for metagenomic sequencing. This is essential for analyzing microbial communities in the water samples.

2.3. Bioinformatics Processing of Metagenomic Sequences

We employed the Illumina sequencing platform for second-generation sequencing of the samples. To ensure data quality, fastp software (V 0.23.4) was utilized for quality control. Read assembly was performed using Megahit's single splicing strategy, and employing the succinct de Bruijn graph method. Cluster analysis was conducted using CD-HIT to identify similarities and differences between samples; cluster similarity and gene sequence clustering coverage parameters were set to greater than 90%. For gene abundance information, the SOAP aligner software (V 1.06) compared the non-redundant gene set. Gene abundance was calculated based on a similarity threshold equal to or higher than 0.95, with insertion fragment lengths between 500 and 300 bp. The non-redundant gene set was compared with the NR database, including bacteria, fungi, archaea and viruses, using DIAMOND (with parameters set to blastp and E-value $\leq 1 \times 10^{-5}$) to obtain biological information from individual samples. This comprehensive approach provided insights into the microbial community structure and potential functions present in the samples. The metagenomic sequencing data from this study have been submitted to the National Center for Biotechnology Information (NCBI) database; this is accessible under accession codes SRR18119958 and SRR18119957.

3. Result and Discussion

3.1. Overall Microbial Community Structure

Microbial community analysis of the Xi River's upstream water samples revealed significant diversity. In total there were 132 phyla, 294 classes, 684 orders, 1252 families, and 3049 genera. Downstream samples, meanwhile, displayed 135 phyla, 297 classes, 680 orders, 1260 families, and 3134 genera. These findings underscore the substantial microbial diversity in both upstream and downstream areas, showing only minor differences in their taxonomic compositions. The detailed taxonomic data derived from DIAMOND analysis provides crucial insights into the diverse microbial communities in the Xi River, highlighting their potential ecological functions in this aquatic environment.

3.2. Alpha Diversity Analysis

We utilized alpha diversity statistical methods to analyze the genetic data obtained from sequencing samples from upstream and downstream of the Xi River. This analysis included calculating the Shannon index at the domain classification level to assess community diversity, and calculating the Shannon evenness index to evaluate community evenness. As illustrated in Figure 2, the results indicate that the microbial community in the upstream samples (XDU) exhibited higher diversity and evenness compared to the downstream samples (XDD) at the domain classification level. The increased diversity and evenness of the upstream microbial community suggest a more balanced and stable ecosystem in that region. These findings further corroborate previous observations that the overall water quality at the upstream sampling points is superior to that at the downstream points. The enhanced diversity and homogeneity of the upstream microbial community imply a healthier and less disturbed ecosystem, while the diversity and homogeneity of the microbial community downstream are reduced due to the influence of the wastewater treatment plant's effluents. A detailed analysis of diversity at finer taxonomic levels will be discussed later in the text.

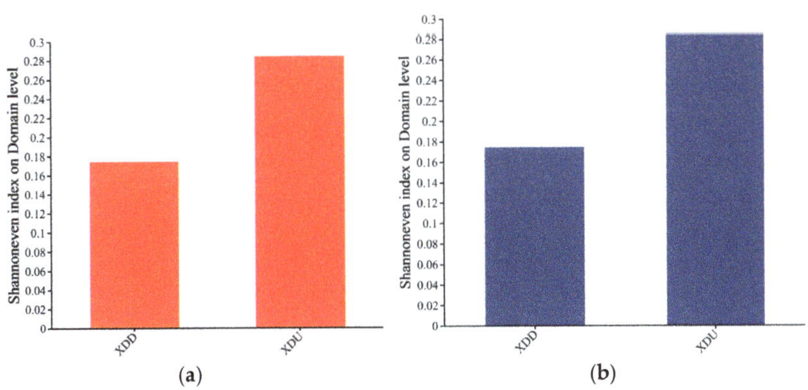

Figure 2. Alpha diversity analysis in upstream (XDU) and downstream (XDD) samples. (**a**) Shannon index; (**b**) Shannon evenness index.

3.3. Taxonomic Analysis at the Phylum and Genus Levels

The metagenomic sampling results, when compared with the NR species database, revealed the abundance of dominant microorganisms. Figure 3a presents these data as a bar chart. The upstream samples (XDU) had six dominant phyla with ≥1% abundance at the phylum level. These included *Proteobacteria* (53.42%), *Bacteroidetes* (19.11%), *Bacillariophyta*

(6.63%), *unclassified_d_Eukaryota* (5.06%), *Cyanobacteria* (3.37%), and *Verrucomicrobia* (1.38%). Conversely, the downstream (XDD) samples showed a broader diversity; there were 13 dominant phyla of ≥1% abundance, highlighting *Proteobacteria* (53.92%), *Bacillariophyta* (2.34%), and others such as *unclassified_d_Bacteria* (2.10%), *Nitrospirae* (1.68%), *Cyanobacteria* (1.52%), *unclassified_d_Eukaryota* (1.36%), *Planctomycetes* (1.33%), and *Gemmatimonadetes* (1.27%).

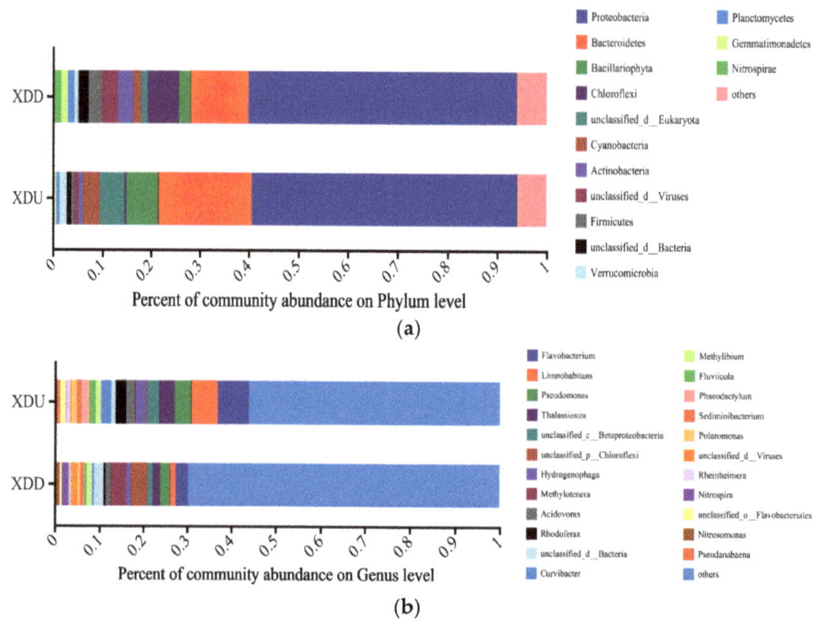

Figure 3. Column chart of microbial communities in upstream (XDU) and downstream (XDD) samples at (**a**) phylum; and (**b**) genus level.

Figure 3b presents the genus-level microbial composition in the Xi River, where the upstream (XDU) sample revealed 3049 genera, 16 of which were predominant genera, i.e., each had over 1% abundance. Among these, *Flavobacterium* (7.12%), *Limnohabitans* (6.09%), and *Pseudomonas* (3.73%) are notable. *Thalassiosira* (3.43%), *Betaproteobacteria* (2.67%), *Hydrogenophaga* (2.59%), and *Rhodoferax* (2.45%) are also significant. Downstream (XDD) samples, however, showed 3134 genera, of which 14 genera were at ≥1% abundance. These included *Chloroflexi* (3.84%), *Methylotenera* (3.40%). *Flavobacterium* (2.80%), *Pseudomonas* (2.24%), *unclassified_c_Bacteria* (2.06%), *Thalassiosira* (1.66%), *Nitrospira* (1.60%), *unclassified_d_Viruses* (1.44%), *Limnohabitans* (1.31%), *Acidovorax* (1.26%), *Betaproteobacteria* (1.25%), *Hydrogenophaga* (1.12%), and *Nitrosomona* (1.08%). The metagenomic data thus provide a comprehensive overview of the microbial ecosystem's complexity and its ecological interplay within the river.

Figure 3 depicts the distribution of microbial taxa at both the phylum and genus levels. At the phylum level, downstream microbes exhibit a more uniform distribution; this is indicative of a diverse microbial presence. This suggests adaptation of various microbial phyla to the downstream environment. This is likely influenced by factors such as wastewater treatment plant discharge. In contrast, at the genus level upstream microbes show a higher degree of evenness, with certain genera dominating within their respective phyla. These dominant genera play key roles in the ecosystem, and contribute to the microbial diversity in the upstream Xi River. These findings highlight the importance of analyzing microbial data across different taxonomic levels, providing a deeper understanding of the ecological structures and the specific dynamics within taxa. Such detailed analysis is essential for assessing the impacts of environmental changes on aquatic ecosystems' microbial communities.

The sequencing analysis in this study focused on the 50 most abundant species, with their relative abundances illustrated on a heatmap using average values. Figure 4 highlights that at the phylum level, downstream microbes have a higher overall relative abundance compared to upstream microbes. This suggests a diverse range of phyla in the downstream environment that are influenced by the wastewater treatment plant effluent and changes in water quality. At the genus level, however, upstream microbes show greater dominance, indicating that specific genera are more prevalent in the upstream region of the Xi River. These dominant genera are likely significant in shaping the ecosystem, and contribute to the observed higher evenness in the upstream microbial community at the genus level.

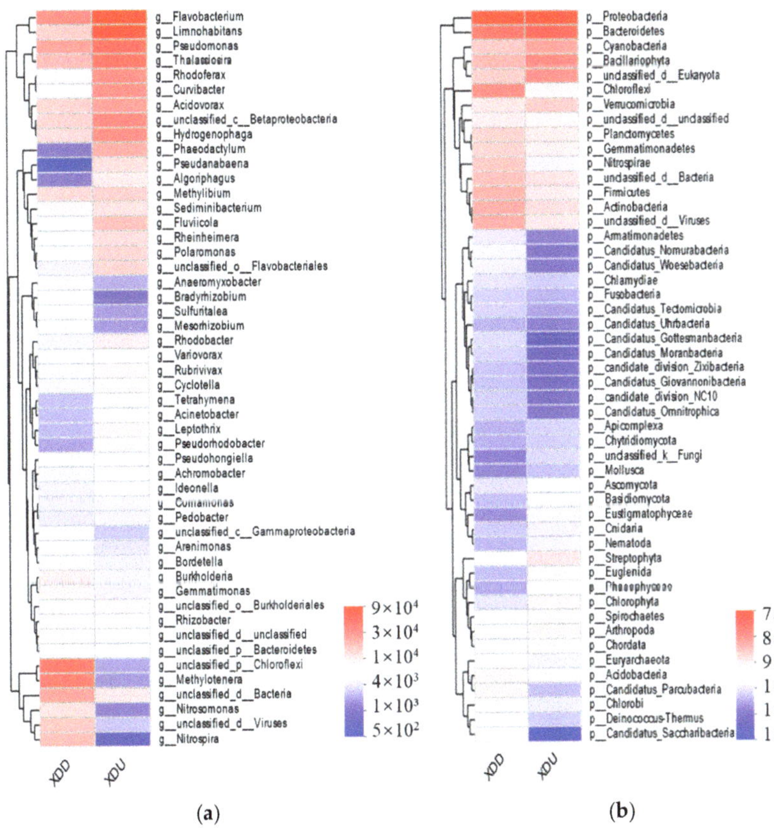

Figure 4. Heatmap of bacterial communities in upstream (XDU) and downstream (XDD) samples at (**a**) genus and (**b**) phylum level.

Observations from Figures 3 and 4 demonstrate the distinct impacts of wastewater treatment plants on the Xi River's microbial community at different taxonomic levels. Downstream, there is a notable increase in the relative abundance of microbial phyla; this indicates a shift toward a more diverse microbial community. In contrast, the upstream region showcases more evenness and dominance at the genus level, suggesting a stable and established microbial structure. This variation in response at different taxonomic levels highlights the complexities in microbial adaptation to environmental shifts and emphasizes the importance of multi-level taxonomic analysis for a comprehensive understanding of the microbial dynamics in the Xi River. Additionally, these findings stress the need for focused monitoring and management approaches to address the specific impacts of wastewater discharges on aquatic microbial communities.

3.4. Kyoto Encylopedia of Genes and Genomes (KEGG) Metabolic Analysis

Microbial populations in the Xi River play a pivotal role in metabolic processes, which are basically linked to the river's physicochemical dynamics. An in-depth examination of these biological functions is essential. The figure shows the classification of metabolic pathways as classified by the Kyoto Encyclopedia of Genes and Genomes (KEGG), focusing on secondary metabolism. This classification allows for a comparative analysis of metabolic variations; statistical tools such as Fisher's exact test and a two-tailed test are used, with a threshold of significance set at a p-value of 0.05.

Figure 5 illustrates that in the Xi River, the overall metabolic pathway's abundance values are higher downstream than upstream, except in most individual metabolic categories, where upstream values prevail. Notably, elevated Xenobiotics biodegradation and metabolism downstream suggest a higher concentration of chemicals, drugs, and organic pollutants in these waters [18]. Additionally, downstream replication and repair values significantly surpass upstream values, indicating more active DNA replication and repair processes downstream. This could signify a greater environmental shift downstream, as reflected by the microbial community's heightened sensitivity and adaptability [19]. Consequently, discharge from the FMWWPT has introduced more pollution, such as organic compounds into the Xi River, thereby imposing stress on downstream microbial populations. These microbes have consequently ramped up their catabolic metabolic functions to cope with the increased organic compound levels. Such adaptability mirrors the microbial response to environmental modifications induced by FMWWPT discharge. In contrast, metabolic aspects such as energy metabolism and translation exhibit higher abundance upstream. This is potentially due to a richer and more diverse microbial community in these regions.

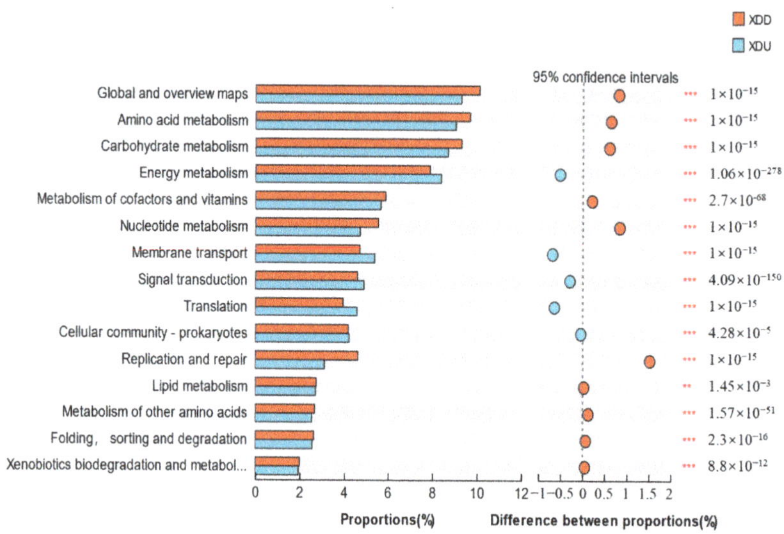

Figure 5. Kyoto Encyclopedia of Genes and Genomes (KEGG) metabolic classification functional differential analysis plot ($0.001 < p \leq 0.01$ **, $p \leq 0.001$ ***).

3.5. Potential Biological Pollution

Activated sludge plays a crucial role in the secondary treatment of wastewater treatment plants [20]. This process is meticulously designed to degrade a variety of organic pollutants [21,22]; this results in the emergence of a highly complex community structure [23,24]. This structure encompasses a diverse ecological network, including prokaryotes, eukaryotes, phages, and various other microorganisms [25]. Following this stage, tertiary treatment focuses on disinfecting and sterilizing the sewage that has undergone secondary treatment. An analysis of the changes in microbial abundance in the Xi River, particularly in relation to activated sludge, reveals the potential biological pollution of the Xi River caused by the FMWWTP. This detailed examination of the microbial abundance changes associated with activated sludge in the Xi River water body emphasizes the potential biological pollution that the FMWWTP might inflict on the Xi River ecosystem.

Among the numerous elements of activated sludge, Proteobacteria emerges as the most prevalent species [26]. Its significant role in the removal of nitrogen and phosphorus, along with the degradation of organic matter, is pivotal. The high abundance of Proteobacteria is crucial for the effective operation of the activated sludge system in wastewater treatment facilities. Moreover, in sewage treatment plants utilizing the AAO (anaerobic anoxic oxic) process, the dominant phyla include *Bacteroidetes, Actinobacteria, Acidobacteria, Chloroflexi, Nitrospirae,* and others [27]. Additionally, it has been observed that the dominant microbial genera in various activated sludge samples demonstrate notable differences in their distribution percentages. This variation underlines the diversity and complexity of microbial communities within activated sludge systems. In the AAO process, for instance, the dominant functional genera comprise *Comamonas, Rhodoferax, Nitrospira, Novosphingobium, Terrimonas, Thauera,* and others, including *Zoogloea, Dechloromonas, Acidovorax* [28–30].

Figure 6a demonstrates that, at the phylum level, apart from a decrease in the downstream abundance of *Bacteroidetes*, the abundances of *Proteobacteria, Actinobacteria, Nitrospirae, Acidobacteria,* and *Chloroflexi* are significantly higher in the downstream samples than in the upstream samples. This observation can be attributed to higher contents of BOD_5, COD, TP, and TN in the downstream samples, which are more closely related to the physical and chemical properties of wastewater treatment plant effluents. Additionally, the presence of corresponding microorganisms in the downstream samples leads to a considerable increase in the abundance of bacterial phyla associated with those microorganisms. The decrease in the abundance of *Bacteroidetes* may be due to its negative correlation with organophosphorus concentration in TP [31,32], suggesting that the FMWWTP discharge could contribute to elevated levels of organic phosphorus in the Xi River's waters.

In Figure 6b, it is evident that the abundance values of *Acidovorax* and *Rhodoferax* at the genus level are lower in the downstream samples compared to the upstream samples. This difference may be attributed to variations in the types of activated sludge employed. However, the downstream samples exhibit higher abundance values of most other bacteria, indicating higher overall microbial activity compared to the upstream samples. Notably, *Zoogloea*, a commonly used component of activated sludge, was observed to have proliferated in the downstream samples [33].

The metagenomic sequencing results revealed the presence of unique downstream microorganisms, including *Pinguiophyceae, Xenacoelomorpha, Poribacteria, Chaetognatha,* and *Nanoarchaeota*. Among these, *Pinguiophyceae* is known to play a significant role in sewage treatment systems [34]. It is speculated that the biological pollution caused by the FMWWTP discharge contributes to the increased abundance of *Pinguiophyceae* in the downstream region.

Analyzing the microbial community structure further shows an increase in the percentage abundance of bacteria and a decrease in the percentage abundance of fungi in the downstream compared to the upstream. Bacteria are commonly found to be the predominant flora in activated sludge [35]. This change is likely a consequence of the chemical contamination of the Xi River due to wastewater treatment plant discharge. These contaminants alter the water's original physicochemical properties, creating a more favorable environment for the growth and reproduction of pollutant-treating bacterial species. As a result, the dominance of bacterial populations increases.

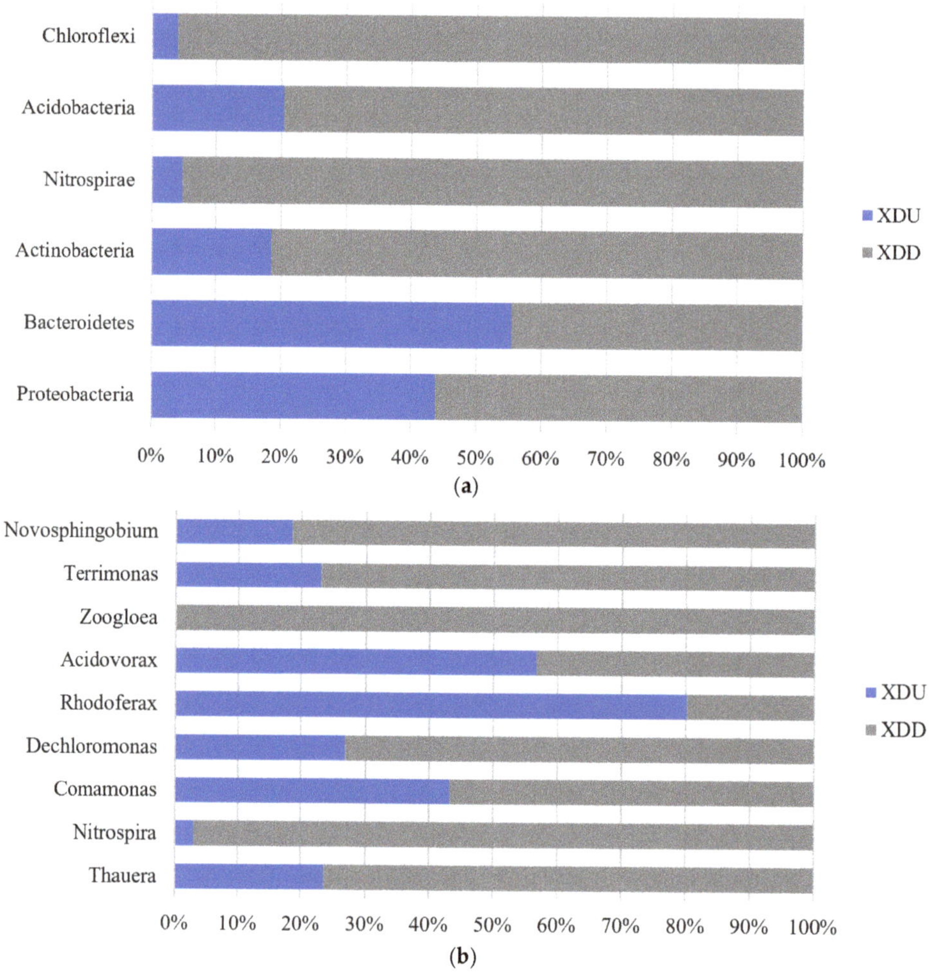

Figure 6. Histograms of bacterial communities (%) in the upstream (XDU) and downstream (XDD) samples at (**a**) phylum and (**b**) genus.

These findings collectively indicate that the FMWWTP has indeed influenced the microbial community structure in the Xi River water body to a certain extent. Moreover, they suggest the potential presence of microbial pollution resulting from wastewater treatment plant discharge. These insights emphasize the need for continuous monitoring and effective management strategies to mitigate the impact of wastewater discharges on the ecological health of the Xi River.

3.6. Water Eutrophication River

Nitrogen and phosphorus are vital elements for the proper functioning of living systems and are essential for human life. However, poorly treated discharge of nitrogen and phosphorus can lead to significant damage to the ecological environment. According to widely accepted international standards, elevated levels of nitrogen and phosphorus in natural water bodies are the primary factors responsible for eutrophication [36]. Eutrophication refers to the excessive growth of algae and aquatic plants in water bodies, which can lead to a cascade of detrimental effects on the ecosystem, such as oxygen depletion, harmful algal blooms, and disruption of the aquatic food chain [37]. Therefore, proper management and treatment of nitrogen and phosphorus discharge from various sources, including wastewater treatment plants, are essential to protect and maintain the ecological balance of water bodies and ensure the sustainability of aquatic ecosystems.

3.6.1. Nitrogen-Content Variations and Their Effects on Microorganisms

A microbial-related denitrification process in water involves three main nitrogen conversion processes: ammonification, nitrification, and denitrification. Nitrogen fixation can occur under both aerobic and anaerobic conditions, while nitrification is specific to aerobic conditions, and denitrification occurs in anoxic conditions [36]. Ammonification refers to the process in which microorganisms decompose organic nitrogen compounds to produce ammonia. Nitrification is the process by which nitrifying bacteria convert ammonia nitrogen in wastewater into nitrite (NO_2^-) and nitrate (NO_3^-). Denitrification is the process where denitrifying bacteria reduce nitrite (NO_2^-) and nitrate (NO_3^-). back to nitrogen gas [37].

As can be inferred from Figure 7, the nitrogen processing activities of microorganisms in the downstream (including ammonification, nitrification, and denitrification) are more active than those upstream. This observation, coupled with the fact that the total nitrogen (TN) content is comparatively lower upstream but the ammonium nitrogen (AN) content is significantly higher upstream than downstream, suggests that the TN levels in the Xi River have risen from upstream to downstream due to the discharge from the FMWWTP; and this subsequently increases the nitrogen processing activities of the microorganisms in the river. Notably, the heightened nitrification activity downstream indicates that during the river's flow, nitrification-related bacteria converted a substantial amount of AN into nitrate nitrogen, leading to a much higher AN content upstream than downstream, as shown in Table 1. Sequencing results also reveal higher typical abundance values of nitrifying bacterial genera, such as *Nitrosococcus* and *Nitrosomonas*, in the downstream samples. These findings collectively demonstrate the significant impact of microbial activity and community composition on the nitrogen transformation processes in the waters of the Xi River. The lower abundance values of the nitrate/nitrite transport system downstream, as shown in Figure 7, may be due to differences in the microbial communities between upstream and downstream, resulting in variations in the associated nitrogen metabolism processes.

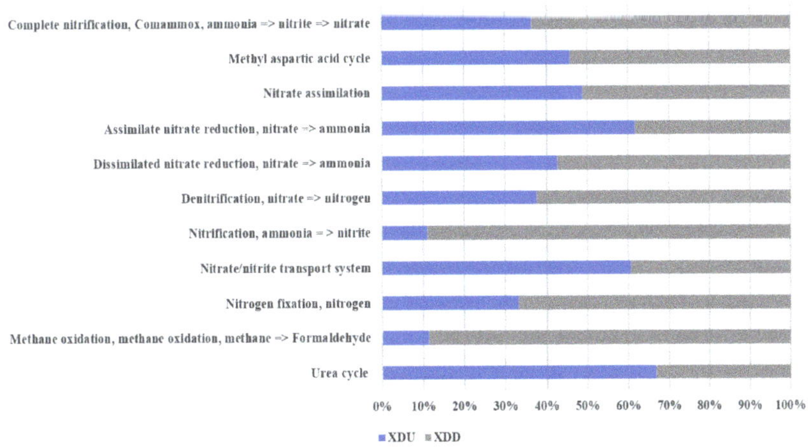

Figure 7. Nitrogen metabolism process of Xi River samples.

3.6.2. Phosphorus-Content Variations and Their Effects on Microorganisms

Polyphosphate-accumulating organisms (PAO), also called phosphorus-absorbing bacteria, are bacteria that degrade large amounts of phosphorus in wastewater under certain environmental factors. Polyphosphate-accumulating organisms are commonly found in ecological environments with high phosphorus content, such as in activated sludge in the enhanced biological phosphate removal (EBPR) system [38]. These organisms are typically found in high-phosphorus ecological environments, such as within the activated sludge of enhanced biological phosphate removal (EBPR) processes. Interestingly, phosphorus-accumulating bacteria can also thrive in some natural water bodies, even those with lower phosphorus levels. In the context of the FM sewage treatment plant, the activated sludge in the AAO process includes phosphorus-accumulating bacteria that efficiently absorb excess phosphorus. The common taxonomic groups of these bacteria encompass *Proteobacteria*, *Bacteroidetes*, *Chloroflexi*, *Planctomycetes*, *Verrucomicrobia*, *Acidobacteria*, among others [39]. This highlights the diversity and adaptability of these bacteria in various environmental settings.

As indicated in Figure 8, among the common polyphosphate-accumulating organisms (PAO), the abundance of related bacteria in the downstream is higher than that in the upstream, with the exception of *Bacteroidetes*. The reduction in *Bacteroidetes* abundance can be associated with the discharge from the FMWWTP. This leads to increased phosphorus and organic phosphorus content in the Xi River. This indicates that the high organic phosphorus content of the discharge has a favorable impact on the survival environment for the majority of polyphosphoric bacteria in the river, thereby promoting their overall growth and reproduction. These shifts in microbial community composition highlight the potential influence of wastewater treatment plant discharge on the ecological dynamics of phosphorus-accumulating bacteria in the Xi River.

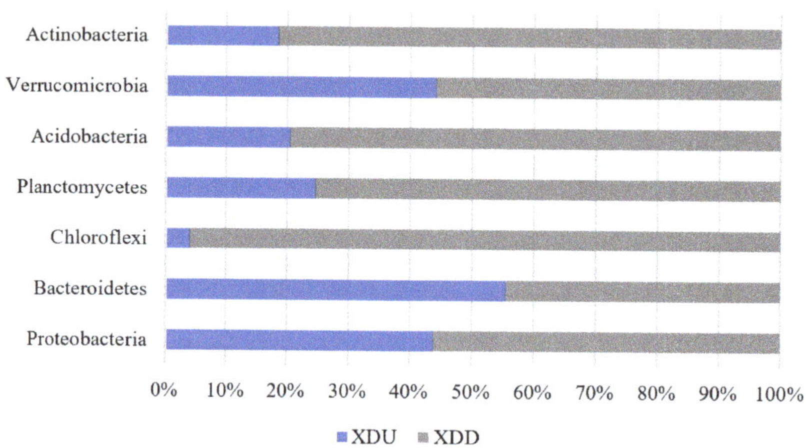

Figure 8. Phosphorus accumulating bacteria at the common phylum level.

3.7. Resistance Gene Pollution

Since the latter half of the 20th century, there has been a substantial increase in the production and usage of antibiotics. Unfortunately, the incomplete absorption of these antibiotics by humans and animals results in significant amounts of antibiotic residues and metabolites being released into the natural environment via excreta. This release exerts selective pressure on microorganisms, and encourages the acquisition of antibiotic resistance genes through mutation and their dissemination through horizontal gene transfer. The enduring presence of these resistance genes in the environment is a significant hurdle in their eradication. Moreover, studies have indicated that resistance genes can transfer from transgenic plants to microorganisms, potentially expanding the spread of antibiotic resistance beyond microbial communities [40]. Elevated antibiotic levels in water environments can interrupt the nitrogen cycle, impeding denitrification and anaerobic ammonia oxidation processes. This disturbance leads to the emission of nitrous oxide, and contributes to environmental problems such as eutrophication, the greenhouse effect, and ozone depletion [41]. Recent studies have highlighted that surface waters receiving wastewater treatment plant effluent show a significant increase in levels of antibiotic-resistant bacteria. This trend indicates that wastewater treatment plant effluents are likely a major source of such bacteria in natural water bodies [42]. The growing presence of antibiotic residues and the escalation of antibiotic-resistant bacteria in aquatic environments underline the critical need for effective waste management, the enhancement of wastewater treatment processes, and judicious use of antibiotics. These measures are vital to protect human health and the environment from the adverse consequences of antibiotic resistance.

As depicted in Figure 9, downstream samples from the Xi River show a heightened abundance compared to upstream samples of various antibiotic resistance gene categories such as sulfonamide, tetracycline, penicillin, streptomycin, and fluoroquinolone. Interestingly, the bacitracin resistance category is more prevalent in upstream samples. This increased presence of resistance genes downstream can be attributed to the influence of the FMWWTP effluent; this is reflected in the observed microbial metabolic profile. The variance in resistance gene abundance between upstream and downstream samples may stem from differences in the disinfection treatment procedures during water discharge and treatment. These procedures might result in varying levels of inactivation of microorganisms carrying these resistance genes [43].

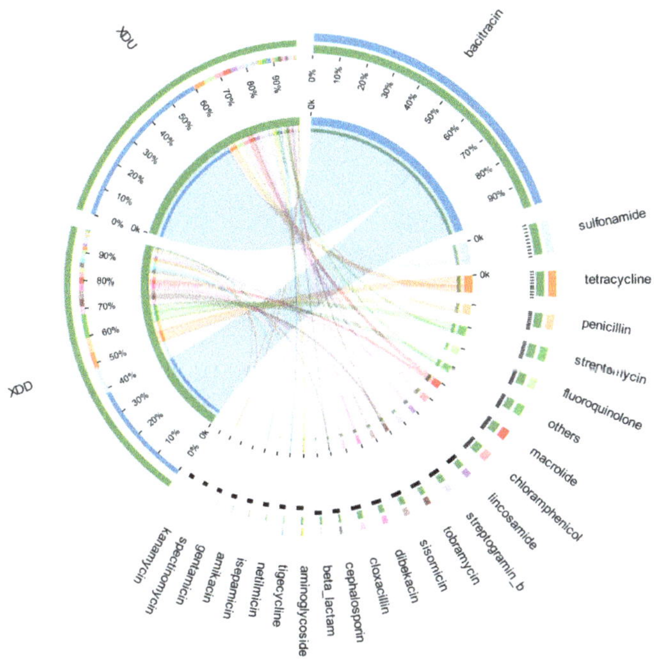

Figure 9. Crossplots of resistance genes upstream (XDU) and downstream (XDD).

In general, the downstream of Xi River is more seriously contaminated with antibiotic resistance genes and needs to be treated effectively. There are various methods for removing antibiotic resistance genes from the river, such as AGS-SBR technology or anaerobic double membrane bioreactor (AnMBR) technology. This can be used in wastewater treatment plants for effective removal of antibiotic resistance genes [44,45]. A variety of modified biochar adsorption, among other methods, can be used for removal of antibiotic resistance genes [46,47]. Additionally, the practice of prudent antibiotic usage is essential to minimize the emergence and dissemination of antibiotic-resistant bacteria and genes.

4. Conclusions

This study comprehensively reveals the interaction between environmental factors and microbial communities in the waters of the Xi River that are impacted by the Fuxin Meng Wastewater Treatment Plant (FMWWTP) discharge. Through metagenomic sequencing and physicochemical property analysis, changes in microbial diversity, metabolic functions, and potential contamination above and below the FMWWTP discharge point were assessed. The results indicate that the microbial community structure of the Xi River is significantly impacted by the pollutants from FMWWTP. Downstream microbial diversity is notably reduced, and there are changes in the dominance of microorganisms or organic matter, as well as an increase in genetic diversity related to metabolic and functional capabilities. These changes may be adaptive responses to stress caused by organic pollutants. Most importantly, the study discovered an increase in the presence of potential biological contaminants and antibiotic resistance genes downstream. This signals risks to environmental quality and public health. The pollution status of the recipient water bodies necessitates improvements in various aspects, such as adopting new wastewater treatment technologies and facilities at the treatment plants, pre-treating non-domestic wastewater, and reducing the misuse of antibiotics in daily life.

The methods used in this study facilitate continuous monitoring of microbial community structure, metabolic functions, and pollution levels in similar environments. This emphasizes the necessity of routine comparisons between downstream and upstream areas. The study advocates for continuous improvements in wastewater treatment processes and adherence to strict environmental standards, and highlights the need for effective management and sustainable practices to protect the ecological integrity of the Xi River and other water bodies affected by human activities.

Author Contributions: Writing—original draft, W.J.; Writing—review & editing, S.S., H.Z. and Y.X.; Visualization, M.Z. All authors have read and agreed to the published version of the manuscript.

Funding: This research was funded by National Key Research & Development Program of China, grant number 2019YFC1803800 and 2019YFC1803801.

Data Availability Statement: Data available in a publicly accessible repository [NCBI, BioProject: PRJNA797902].

Conflicts of Interest: The authors declare no conflict of interest.

References

1. Alharbi, O.A.; Jarvis, E.; Galani, A.; Thomaidis, N.S.; Nika, M.-C.; Chapman, D.V. Assessment of selected pharmaceuticals in Riyadh wastewater treatment plants, Saudi Arabia: Mass loadings, seasonal variations, removal efficiency and environmental risk. *Sci. Total Environ.* **2023**, *882*, 163284. [CrossRef]
2. Hernandez-Chover, V.; Castellet-Viciano, L.; Fuentes, R.; Hernandez-Sancho, F. Circular economy and efficiency to ensure the sustainability in the wastewater treatment plants. *J. Clean. Prod.* **2023**, *384*, 135563. [CrossRef]
3. Li, Y.; Han, Q.; Li, B. Engineering-scale application of sulfur-driven autotrophic denitrification wetland for advanced treatment of municipal tailwater. *Bioresour. Technol.* **2023**, *379*, 129035. [CrossRef] [PubMed]
4. Li, J.; Tao, T.; Li, X.-B.; Li, J.-H. A pilot-scale experimental research on biological phosphorus removal in a modified SBR treating urban wastewater. In Proceedings of the 2nd International Conference on Chemical Engineering and Advanced Materials (CEAM 2012), Guangzhou, China, 13–15 July 2012; pp. 2329–2332.
5. Douxfils, J.; Mandiki, R.; Silvestre, F.; Bertrand, A.; Leroy, D.; Thome, J.-P.; Kestemont, P. Do sewage treatment plant discharges substantially impair fish reproduction in polluted rivers? *Sci. Total Environ.* **2011**, *409*, 4139. [CrossRef]
6. Zhu, T.-T.; Su, Z.-X.; Lai, W.-X.; Zhang, Y.-B.; Liu, Y.-W. Insights into the fate and removal of antibiotics and antibiotic resistance genes using biological wastewater treatment technology. *Sci. Total Environ.* **2021**, *776*, 145906. [CrossRef]
7. Wang, J.; Chen, Y.; Cai, P.; Gao, Q.; Zhong, H.; Sun, W.; Chen, Q. Impacts of municipal wastewater treatment plant discharge on microbial community structure and function of the receiving river in Northwest Tibetan Plateau. *J. Hazard. Mater.* **2022**, *423*, 127170. [CrossRef]
8. Igere, B.E.; Okoh, A.I.; Nwodo, U.U. Wastewater treatment plants and release: The vase of Odin for emerging bacterial contaminants, resistance and determinant of environmental wellness. *Emerg. Contam.* **2020**, *6*, 212–224. [CrossRef]
9. Wommack, K.E.; Ravel, J. *Microbiome*, demystifying the role of microbial communities in the biosphere. *Microbiome* **2013**, *1*, 1. [CrossRef]

10. Shade, A. Diversity is the question, not the answer. *ISME J.* **2017**, *11*, 1–6. [CrossRef]
11. Yang, H.; Xu, M.; Wang, L.; Wang, X.; Jeppesen, E.; Zhang, W. Metagenomic analysis to determine the characteristics of antibiotic resistance genes in typical antibiotic-contaminated sediments. *J. Environ. Sci.* **2023**, *128*, 12–25. [CrossRef]
12. Li, M.; Zhao, S.; Zheng, N.; Wang, J. Advances in Microbial Macrogenomic Data Analysis Methods. *J. Mirobiol.* **2020**, *40*, 89–95.
13. Chen, G.; Bai, R.; Zhang, Y.; Zhao, B.; Xiao, Y. Application of metagenomics to biological wastewater treatment. *Sci. Total Environ.* **2022**, *807*, 150737. [CrossRef]
14. *GB 3838-2002*; Environmental Quality Standards for Surface Water. Chinese Research Academy of Environmental Sciences: Beijing, China, 2002.
15. Zhu, H.; Li, B.; Ding, N.; Hua, Z.; Jiang, X. A Case Study on Microbial Diversity Impacts of a Wastewater Treatment Plant to the Receiving River. *J. Geosci. Environ. Prot.* **2021**, *9*, 206–220. [CrossRef]
16. Zhang, X.; Zhang, X.-Q.; Yu, H.-B.; Song, H.-L.; Du, M.-X. Phosphorus Removal from Wastewater by Electrocoagulation with Magnetized Iron Particle Anode. *Water Air Soil Pollut.* **2020**, *231*, 502. [CrossRef]
17. Wang, Q.L.C. Quantitative assessment study of water pollution composite index in Fuxin Xi River Basin. *Heilongjiang Water Sci. Technol.* **2021**, *49*, 17–19+47. [CrossRef]
18. Chao, Y.; Ma, L.; Yang, Y.; Ju, F.; Zhang, X.-X.; Wu, W.-M.; Zhang, T. Metagenomic analysis reveals significant changes of microbial compositions and protective functions during drinking water treatment. *Sci. Rep.* **2013**, *3*, 3550. [CrossRef]
19. Feng, M.; Lu, Y.; Yang, Y.; Zhang, M.; Xu, Y.-J.; Gao, H.-L.; Dong, L.; Xu, W.-P.; Yu, S.-H. Bioinspired greigite magnetic nanocrystals: Chemical synthesis and biomedicine applications. *Sci. Rep.* **2013**, *3*, 2994. [CrossRef]
20. Bae, H.-K. A Comparative Analysis of Two Different Wastewater Treatment Processes in Actual Wastewater Treatment Plants. *Quant. Bio-Sci.* **2018**, *37*, 19–26. [CrossRef]
21. Alkan, U.; Eleren, S.C.; Nalbur, B.E.; Odabas, E. Influence of the activated sludge system configuration on heavy metal toxicity reduction. *World J. Microbiol. Biotechnol.* **2008**, *24*, 1435–1443. [CrossRef]
22. Ouyang, J.; Li, C.; Wei, L.; Wei, D.; Zhao, M.; Zhao, Z.; Zhang, J.; Chang, C.-C. Activated sludge and other aerobic suspended culture processes. *Water Environ. Res.* **2020**, *92*, 1717–1725. [CrossRef]
23. Ya, X.; Jingcai, L.; Lu, D.; Yuqiang, L.; Weishi, L.; Changxing, N.; Qifei, H. Buffering distance between hazardous waste landfill and water supply wells in a shallow aquifer. *J. Clean. Prod.* **2019**, *211*, 1180–1189. [CrossRef]
24. Yu, K.; Zhang, T. Metagenomic and Metatranscriptomic Analysis of Microbial Community Structure and Gene Expression of Activated Sludge. *PLoS ONE* **2020**, *15*, e0243233. [CrossRef]
25. Wang, F.; Sun, X.; Li, J.; Ai, S.; Ren, Q.; Nie, Z.; Bian, D. Analysis of microbial characteristics and population difference between different compartments of the mixed system at low temperature. *Desalination Water Treat.* **2021**, *225*, 53–62. [CrossRef]
26. Zhang, L.; Shen, Z.; Fang, W.; Gao, G. Composition of bacterial communities in municipal wastewater treatment plant. *Sci. Total Environ.* **2019**, *689*, 1181–1191. [CrossRef]
27. Ma, Q.-Q.; Yuan, L.-J.; Niu, Z.-D.; Zhao, J.; Huang, C. Microbial Community Structure of Activated Sludge and its Response to Environmental Factors. *Huan Jing Ke Xue Huanjing Kexue* **2021**, *42*, 3886–3893. [CrossRef]
28. Lou, T.; Peng, Z.; Jiang, K.; Niu, N.; Wang, J.; Liu, A. Nitrogen removal characteristics of biofilms in each area of a full-scale AAO oxidation ditch process. *Chemosphere* **2022**, *302*, 134871. [CrossRef]
29. Liu, F.; Qian, G.; Zhao, X.; Hu, X. Mechanism insights into the nitrogen removal of bio augmented AAO: Specifically focusing on the nitrogen metabolic pathways and microbial taxa-functional genes associations. *J. Water Process Eng.* **2022**, *50*, 103245. [CrossRef]
30. Wang, L.; Yuan, L.; Li, Z.-H.; Zhang, X.; Leung, K.M.Y.; Sheng, G.-P. Extracellular polymeric substances (EPS) associated extracellular antibiotic resistance genes in activated sludge along the AAO process: Distribution and microbial secretors. *Sci. Total Environ.* **2022**, *816*, 151575. [CrossRef]
31. Yan, X.; He, M.; Zheng, J.; Zhu, T.; Zou, Z.; Tang, B.; Yu, Y.; Mai, B. Tris (1,3-dichloro-2-propyl) phosphate exposure disrupts the gut microbiome and its associated metabolites in mice. *Environ. Int.* **2021**, *146*, 106256. [CrossRef]
32. Wang, D.; Yan, S.; Yan, J.; Teng, M.; Meng, Z.; Li, R.; Zhou, Z.; Zhu, W. Effects of triphenyl phosphate exposure during fetal development on obesity and metabolic dysfunctions in adult mice: Impaired lipid metabolism and intestinal dysbiosis. *Environ. Pollut.* **2019**, *246*, 630–638. [CrossRef]
33. An, W.; Guo, F.; Song, Y.; Gao, N.; Bai, S.; Dai, J.; Wei, H.; Zhang, L.; Yu, D.; Xia, M.; et al. Comparative genomics analyses on EPS biosynthesis genes required for floc formation of Zoogloea resiniphila and other activated sludge bacteria. *Water Res.* **2016**, *102*, 494–504. [CrossRef]
34. McMahon, K.D.; Jenkins, D.; Keasling, J.D. Polyphosphate kinase genes from activated sludge carrying out enhanced biological phosphorus removal. *Water Sci. Technol.* **2002**, *46*, 155–162. [CrossRef]
35. Chen, Y.; Geng, N.; Hu, T.; Baeyens, J.; Wang, S.; Su, H. Adaptive regulation of activated sludge's core functional flora based on granular internal spatial microenvironment. *J. Environ. Manag.* **2022**, *319*, 115714. [CrossRef]
36. Ding, S.; Dan, S.F.; Liu, Y.; He, J.; Zhu, D.; Jiao, L. Importance of ammonia nitrogen potentially released from sediments to the development of eutrophication in a plateau lake. *Environ. Pollut.* **2022**, *305*, 119275. [CrossRef]
37. Smith, V.H.; Schindler, D.W. Eutrophication science: Where do we go from here? *Trends Ecol. Evol.* **2009**, *24*, 201–207. [CrossRef]
38. He, S.; Gall, D.L.; McMahon, K.D. "Candidatus accumulibacter" population structure in enhanced biological phosphorus removal Sludges as revealed by polyphosphate kinase genes. *Appl. Environ. Microbiol.* **2007**, *73*, 5865–5874. [CrossRef]

39. Oehmen, A.; Lemos, P.C.; Carvalho, G.; Yuan, Z.; Keller, J.; Blackall, L.L.; Reis, M.A.M. Advances in enhanced biological phosphorus removal: From micro to macro scale. *Water Res.* **2007**, *41*, 2271–2300. [CrossRef]
40. Pontiroli, A.; Rizzi, A.; Simonet, P.; Daffonchio, D.; Vogel, T.M.; Monier, J.-M. Visual Evidence of Horizontal Gene Transfer between Plants and Bacteria in the Phytosphere of Transplastomic Tobacco. *Appl. Environ. Microbiol.* **2009**, *75*, 3314–3322. [CrossRef] [PubMed]
41. Chee-Sanford, J.C.; Mackie, R.I.; Koike, S.; Krapac, I.G.; Lin, Y.-F.; Yannarell, A.C.; Maxwell, S.; Aminov, R.I. Fate and Transport of Antibiotic Residues and Antibiotic Resistance Genes following Land Application of Manure Waste. *J. Environ. Qual.* **2009**, *38*, 1086–1108. [CrossRef] [PubMed]
42. Galvin, S.; Boyle, F.; Hickey, P.; Vellinga, A.; Morris, D.; Cormican, M. Enumeration and Characterization of Antimicrobial-Resistant Escherichia coli Bacteria in Effluent from Municipal, Hospital, and Secondary Treatment Facility Sources. *Appl. Environ. Microbiol.* **2010**, *76*, 4772–4779. [CrossRef] [PubMed]
43. Munir, M.; Wong, K.; Xagoraraki, I. Release of antibiotic resistant bacteria and genes in the effluent and biosolids of five wastewater utilities in Michigan. *Water Res.* **2011**, *45*, 681–693. [CrossRef] [PubMed]
44. Perez-Bou, L.; Gonzalez-Martinez, A.; Gonzalez-Lopez, J.; Correa-Galeote, D. Promising bioprocesses for the efficient removal of antibiotics and antibiotic-resistance genes from urban and hospital wastewaters: Potentialities of aerobic granular systems. *Environ. Pollut.* **2024**, *342*, 123115. [CrossRef]
45. Wang, K.; Zhou, L.; Meng, S.; Wang, Y.; Yu, B.; Wang, J. Anaerobic membrane bioreactor for real antibiotic pharmaceutical wastewater treatment: Positive effect of fouling layer on antibiotics and antibiotic resistance genes removals. *J. Clean. Prod.* **2023**, *409*, 137234. [CrossRef]
46. Ding, Z.; Hu, X.; Wan, Y.; Wang, S.; Gao, B. Removal of lead, copper, cadmium, zinc, and nickel from aqueous solutions by alkali-modified biochar: Batch and column tests. *J. Ind. Eng. Chem.* **2016**, *33*, 239–245. [CrossRef]
47. Zhou, Y.; Leong, S.Y.; Li, Q. Modified biochar for removal of antibiotics and antibiotic resistance genes in the aqueous environment: A review. *J. Water Process Eng.* **2023**, *55*, 104222. [CrossRef]

water

MDPI

Article

The First Evidence of the Water Bioremediation Potential of *Ficopomatus enigmaticus* (Fauvel 1923): From Threat to Resource?

Manuela Piccardo [1,*], Verdiana Vellani [1], Serena Anselmi [2,3], Tecla Bentivoglio [2], Francesca Provenza [2], Monia Renzi [1,3] and Stanislao Bevilacqua [1,3]

1 Department of Life Sciences (DSV), University of Trieste, Via L. Giorgieri 10, 34127 Trieste, Italy; verdiana.vellani@phd.units.it (V.V.); mrenzi@units.it (M.R.); sbevilacqua@units.it (S.B.)
2 Bioscience Research Center, Via Aurelia Vecchia 32, 58015 Orbetello, Italy; serena.anselmi@bsrc.it (S.A.); tecla.bentivoglio@bsrc.it (T.B.); francesca.provenza@bsrc.it (F.P.)
3 Consorzio Nazionale Interuniversitario per le Scienze del Mare (CoNISMa), Piazzale Flaminio, 9, 00196 Roma, Italy
* Correspondence: manuela.piccardo@units.it

Abstract: Each year, a staggering 700,000 tons of synthetic dyes are manufactured globally, leading to the release of dye-laden wastewater into aquatic systems. These synthetic dyes resist biodegradation, endangering human and environmental health. Since traditional wastewater treatments are basically unable to remove dyes, exploring the potential of alternative solutions, such as bioremediation, is crucial to reduce dye contamination in aquatic ecosystems. *Ficopomatus enigmaticus* (Fauvel 1923), listed as one of the 100 worst invasive species in Europe, is considered an invasive ecosystem engineer capable of causing economic and ecological losses. Despite this negative status, the literature suggests its positive contributions to aquatic ecosystems as habitat former and water bioremediator. However, existing evidence on the potential of *F. enigmaticus* to improve water quality is fragmented and lacks experimental data from laboratory tests. This study examined the potential of *Ficopomatus* reefs, both living and dead, to enhance water quality by removing contaminants, focusing on methylene blue (MB), one of the most common synthetic dyes. Bioaccumulation and bioadsorption were identified as key mechanisms for dye removal, supported by ATR-FTIR and microscopic analyses. *Ficopomatus* efficiently removed up to 80% of MB within 24 h. Bioaccumulation in the soft body accounted for 18% of the total removal, while complex adsorption phenomena involving carbonaceous, microalgal, and organic reef components accounted for 82%. Surprisingly, bioremediated solutions exhibited significant effects in ecotoxicological tests on bacteria, indicating the potential of *F. enigmaticus* to disrupt bacterial quorum sensing related to biofilm formation, and suggesting a possible antifouling action. This study underscores the intricate interplay between *F. enigmaticus*, water quality improvement, and potential ecological consequences, stressing the need for further investigation into its multifaceted role in aquatic ecosystems.

Keywords: bioremediation; water purification; methylene blue; synthetic dye; bioadsorption; bioaccumulation; ecosystem service; invasive species; antifouling; toxicity

1. Introduction

Ficopomatus enigmaticus (Fauvel, 1923) is a tube-building polychaete worm belonging to the family Serpulidae. The origin of *F. enigmaticus* still remains unknown, although most probably, it is native of temperate regions of the Indian Ocean [1]. Today, *F. enigmaticus* is found worldwide, between the northern hemisphere isotherm of 16 °C and the 21 °C isotherm of the southern hemisphere [2], preferentially inhabiting coastal brackish water lagoons at temperate latitudes. *F. enigmaticus* was discovered in 1921 in northern France by Fauvel (1923) [3] and it is listed as one of the 100 worst invasive species in Europe [4]. However, there are currently no unified strategies for its control or management. This species exhibits a fast growth rate and high tolerance to variable environmental conditions, leading to significant ecological impacts. It hosts a pollution-resistant invertebrate community [5] and alters the surrounding environmental conditions by modifying water flow regimes, affecting sediment grain size [6], and enriching the sediment with organic carbon and nitrogen compounds through the release of *feces* and *pseudofeces* [7].

The proliferation of this species may also result in economic losses as it can negatively impact ships, buoys, and harbor structures [1] and build extensive reefs that can physically impede navigation and hinder the deployment of fishnets [8]. There is an ongoing debate about the invasiveness or naturalization of this species and, despite its putative negative effects, *F. enigmaticus* has also been suggested to enhance the status of aquatic ecosystems, acting as a habitat former. Indeed, serpulid reefs are generally considered a biodiversity hotspot capable of supporting a higher abundance and diversity of associated fauna [9–11]. For example, *Ficopomatus* reefs not only provide refuge, substrate, and feeding ground to a number of aquatic invertebrates, but may also serve as resting sites for some birds, such as swans [12]. A further effect of the presence of large populations of *F. enigmaticus* may concern the improvement of water quality by reducing the organic and nutrient load of water bodies as a result of their filter-feeding activity [13,14]. However, the potential role of this species as a bioremediator has been more advocated rather than actually documented.

In this study, experiments in aquaria were carried out to explore the ability of *Ficopomatus* reefs (both living reefs and their dead biogenic structures) to enhance water quality and remove harmful substances from the aquatic medium. We focused on the synthetic dye methylene blue (MB) as an emblematic example of aquatic contaminants. Every year, about 7×10^5 tons of synthetic dyes are produced worldwide, the majority of them being currently used in the textile industry, totaling approximately ten thousand varieties [15–17]. Dyes deteriorate the environmental quality of water bodies by increasing the biochemical and chemical oxygen demand, hindering photosynthesis, inhibiting plant growth, entering the food chain, displaying resistance and bioaccumulation, and potentially contributing to toxicity, mutagenicity, and carcinogenicity in aquatic organisms [17–19]. MB is one of the most used substances in the dye industry, which is commonly used to dye silk, wool, cotton, and paper [18]. Moreover, the food, cosmetics, and pharmaceutical industries also contribute significantly to the use of MB [15,20,21]. Given its extensive industrial usage, a substantial volume of wastewater containing MB is discharged into groundwater and surface water (including brackish, marine, and freshwater bodies), since conventional water treatment processes are typically unable to remove dye contamination [22]. Once released into the environment, MB and dyes in general can persist for a long time due to their high stability against degradation from temperature, light, and water actions, posing severe threats to aquatic environments and human health [15].

We also investigated the possible mechanisms underlying the process of MB decontamination mediated by *Ficopomatus* reefs, including both bioaccumulation and bioadsorption, and tested the toxicity of the bioremediated aquatic medium to understand whether *Ficopomatus* reefs may effectively remove and detoxify cationic dyes, as well as to shed light on their potential contribution to water purification functions in aquatic ecosystems.

2. Materials and Methods

2.1. Animal Sampling and Acclimatization

Ficopomatus enigmaticus was collected in May 2023 within a brackish coastal lake near Koper (Slovenia) recognized as having an established population of this species (45°32′29″ N 13°43′24″ E [23]). Once in the laboratory, the animals were kept in 6 L aquaria filled with 0.45 μm filtered natural seawater at reduced salinity (20 ppt, f-NSW-20; pH 8.0 ± 0.1) and maintained under constant environmental conditions conducive to the well-being of the animal (temperature: 20 °C; light: 12 h/12 h; weekly feeding with *Isochrysis galbana*) for 40 days. During this depuration period, the water was changed weekly and the associated vagile fauna were carefully removed to prevent potential interference in the experiments.

One day (24 h) before the start of the experiment, the main reef was divided into small blocks, each containing approximately 60 individuals (corresponding to 13.22 ± 1.12 g of wet weight, mean ± standard deviation), in analogy to other ecotoxicological studies focusing on *F. enigmaticus* adults (e.g., Ref. [24] used 50 g of reef per 1 L of solution, which corresponds to the 12–13 g of reef per 250 mL used in this study). These blocks were transferred to 250 mL glass beakers for a 24 h acclimatization under the same experimental conditions described earlier (20 °C; 12:12 light/dark; pH 8.0 ± 0.1; continuous stirring at 200 rpm). No food was provided 48 h before or during the experiment.

2.2. *Methylene Blue Solution*

The tests were performed with an MB (No. M 9140, Sigma-Aldrich, St Louis, MO, USA) solution prepared with f-NSW-20. MB is a well-known primary thiazine, a highly water-soluble cationic dye [18]. MB has a characteristic deep-blue color when oxidized and it is colorless in the reduced form [25]. Therefore, the color attenuation of the MB water solution in the experimental treatments, expressed in terms of the reduced optical density (OD), can be used as an indirect measure of the MB removal capacity of the tested organisms [26,27]. The maximum absorbance of the MB solution was investigated by scanning different sample solutions (0.5–5–10–15–20–25 mg/L) between 400 and 800 nm using a UV–visible spectrophotometer (DU 730, Beckman Coulter, Brea, CA, USA). In agreement with previous evidence [26,27], the maximum absorbance value for each concentration is achieved at λ = 660 nm, which was the value used to perform the experiment.

2.3. *Bioremediation Experiment*

In the experiments, both living and dead reefs of *F. enigmaticus* were used for the comparison of their bioremediation capacity in order to disentangle the contribution of bioaccumulation associated with the soft body of the animals and their biogenic hard structures. Concentrations of MB ranging from 50 to 200 mg/L are generally well tolerated by aquatic organisms in analogous experiments [26–28]. However, since the potential toxicological effects of these concentrations on *F. enigmaticus* are unknown, a precautionary approach was preferred and the test concentration was reduced to 25 mg/L, with a total exposure time of 24 h, to prevent excessive stress for the animals.

For the experiment on living reefs, eight small blocks of *F. enigmaticus* (approximately 460 individuals in total) were randomly selected and submerged in 200 mL of a filtered seawater solution of MB at a concentration of 25 mg/L in separate 250 mL glass beakers. The samples were kept in constant agitation (200 rpm) for 24 h at 20 °C. For the experiment on dead reefs, since the physical removal of the worms from their tubes would have damaged the bioconstruction, dead reefs were obtained by drying the living reef blocks for 20 h at a constant temperature of 40 °C to kill the worms while leaving untouched their biogenic structure. Five random replicate blocks of dead reefs were then used for the experimental exposure to MB following the exact same protocol as for the living reefs.

Methylene blue is a widely recognized cationic thiazine dye that is highly water-soluble and exhibits a deep-blue color when solubilized. Consequently, it is customary in the literature to employ color attenuation, expressed in terms of optical density (OD), as an indicator of methylene blue removal capacity [26,27]. Accordingly, the ability of *F. enigmaticus* to remove MB from the aquatic medium was evaluated spectrophotometrically using a UV/Vis spectrophotometer (DU 730 Beckman Coulter) by monitoring, at regular intervals (0–1–2–4–24 h), the decrease in the optical density of the water at a wavelength of 660 nm. In each time of sampling, three replicates of the treatment solution of 3 mL each were sampled in the center of the beaker with constant stirring. After the completion of the spectrophotometric measurements, the samples were returned to their respective beakers to prevent an excessive drop in the water level during the whole experiment.

A negative control with living reef blocks in beakers filled with filtered natural seawater at 20 ppt salinity (hereafter referred to as CN, with five replicates) was implemented. This control was used to evaluate possible spectral interferences due to the release of mucus under the experimental conditions and to correct, through subtraction, the optical density (OD) of the MB-treated samples. A second positive control, consisting of beakers filled with 25 mg/L of the MB seawater solution without animals (hereafter referred to as CP, with four replicates) was used to evaluate the contribution of the possible light photo-degradation process [27] and the possible changes in the solubility of MB in a solution with a high ionic load (i.e., seawater at 20 ppt salinity).

The adsorbent dose, pH, initial MB concentration, and temperature are known to be operating conditions and parameters that can influence the uptake of MB on various adsorbents [15]. For these reasons, the mass of the *Ficopomatus* reef, the initial concentration of the contaminants, and the temperature were maintained constant over the 24 h of exposure (i.e., 13.22 ± 1.12 g of wet weight, 25 mg/L, and 20 °C, respectively). The pH of the dye solutions was checked at the beginning and end of the experiments; a slight average decrease in pH (<0.5 units) was recorded at the end of the experiments for all treatments (Table S1 in Supplementary Material).

2.4. Expression of the Bioremediation Performance

The ability of *F. enigmaticus* to remove MB was assessed by calculating both the decolorization percentage ($D\%$) and the quantity of dye eliminated per unit of biomass at time t (qt).

The decolorization percentage was calculated using Equation (1) [19] as follows:

$$Decolorization\ \% = D\% = \left(\frac{A_{CP} - A_S}{A_{CP}}\right) \times 100 \tag{1}$$

where A_{CP} is equal to the absorbance at t hours of the positive control; A_S is the absorbance at t hours of the samples (living or dead); and both corrected by the negative control (CN) (i.e., by subtracting the absorbance of CN from the absorbance of A_{CP} and A_S).

As the decolorization rate expressed in Equation (1) may be affected by the volume of water and grams of biomass present in the experimental unit, we also calculated the amount of dye eliminated per unit of biomass at time t using the Formula (2) [29] as follows:

$$qt = \left(\frac{C_{CP} - C_S}{m}\right) \times V \tag{2}$$

where V is the volume of water in the beaker (200 mL); m is the wet weight in grams of reef (both for living and dead); C_{CP} is the positive control concentration after t hours; and C_S is the MB concentration after t hours. The optical density was converted to MB concentrations using an equation obtained by fitting a linear regression of MB concentrations against OD (ranging from 0.5 to 25 mg/L, $R^2 = 0.978$).

2.5. Reef Surface Characterization

The reef surface was observed under a stereomicroscope (Nikon—P-DSL32, Landsberg, Germany), and high-definition images were captured with a digital camera (Nikon—DSFi3, Landsberg, Germany). Since a strong MB adsorption was observed, the surface of the reefs was analyzed by means of attenuated total reflectance (ATR) Fourier-transform infrared spectroscopy (FTIR), and collected spectra were matched with reference materials in order to identify the components likely involved in the adsorption process. The infrared spectra were recorded within the 3800–800 cm^{-1} range using an FTIR spectrometer from ThermoFisher—iN10 MX (Waltham, MA, USA). The ATR-FTIR analysis was carried out only on dead reefs. Tube samples from a single, randomly selected reef were pulverized with the aid of an agate pestle and mortar and placed on glass supports for the acquisition of the spectrum via ATR analysis with germanium iridium crystals. The spectra obtained were analyzed using the OMINC Software (version 8.2.0.403), applying components through the matching of libraries acquired from ThermoFisher and internally generated at the BsRC laboratory.

2.6. Toxicity Test

The toxicity of the solutions bioremediated by living and dead *Ficopomatus* reefs was assessed using a bioluminescence inhibition assay that employed the marine Gram-negative bacterium *Aliivibrio fischeri* as the test species. *A. fischeri* is known to be sensitive to a wide range of toxicants and is used in standard acute bioassays [30–32]. Light production is directly proportional to the metabolic activity of the bacterial population, and any inhibition of enzymatic activity due to toxicity or cell death results in a corresponding decrease in the bioluminescence produced by the colony.

Test samples were exposed to freshly prepared bacteria (NRRL B-11177 strain and vial from internal propagation) for a predetermined period (15 and 30 min), and the degree of light emission inhibition was compared to a negative control. This assay provided a measure of both the sub-lethal response and lethality through the degree of inhibition of bioluminescence.

More specifically, a standard Microtox® acute assay was performed following a procedure that is sensitive to changes in bioluminescence and yields highly reproducible results (UNI EN ISO 11348-2:2019 [32]). Briefly, MB and bioremediated MB solutions were tested alongside positive controls (3.5- Dichlorophenol) and negative controls (ASW, artificial sea water, UNI EN ISO 11348-2:2019 [32]. Each treatment was run in duplicate, and the relative bioluminescence was measured using a Microtox® Model 500 Analyzer (ModernWater, 5003054, New Castle, DE, USA) after an incubation period of 15–30 min. Bioluminescence inhibition was expressed as the percentage of the initial concentration reduction in luminescence compared to the luminescence intensity of the control sample.

2.7. Statistical Analysis

A statistical analysis was performed using the GraphPad Prism software (GraphPad Software version 6, San Diego, CA, USA,). Data on the bioremediation performance are represented as mean ± standard deviation. The normality of the data was previously examined using a D'Agostino and Pearson tests, and the homoscedasticity was checked. Consequently, the statistical significance ($\alpha = 0.05$) of the comparisons between living and dead reefs was studied using parametric or nonparametric (Mann–Whitney) unpaired *t*-tests depending on the case. Data on the bioassay are expressed as mean ± standard deviation.

3. Results

3.1. Bioremediation Experiment

MB was partially removed by both living and dead reefs. The changes in color of the solutions, the reef structure, and the naked worm between the start ($t = 0$) and after 24 h of exposure were clearly visible for both the living reef (Figure 1) the dead reef samples (Figure 2).

For both the living and dead reefs, the percentage of removal ($D\%$) increased over time (Figure 3). The living reefs were capable of removing 32.75% of the dye in the first hour, with the maximum percentage of decolorization, 80.43%, achieved after 24 h (Table 1). The dead reefs were less efficient, reaching a maximum percentage of dye removal (64.61%) after 24 h. The performance of the living and dead reefs was significantly different at all times ($p < 0.05$), with the largest differences in decolorization observed at 1 h (17.70%, $p = 0.0016$).

The capacity of dye removal per unit of biomass at time t (qt) reached 0.53 mg/g and 0.31 mg/g after 24 h for the living and dead reefs, respectively (Table 1). The performance of the living and dead reefs was always statistically different ($p < 0.05$), with the largest difference in dye removal per unit of biomass observed at 24 h (0.22 mg/g, $p = 0.0016$).

Figure 1. Change in the color of the MB seawater solutions (top pair of photos), living *Ficopomatus* reef blocks (middle pair of photos), and the naked bodies of the worms (bottom pair of photos) at the start of the experiment (0 h) and at the end of the experiment (after 24 h).

Table 1. Decolorization percentage (*D*%; mean ± standard deviation) and dye removal capacity (*qt*; mean ± standard deviation) of living and dead reefs of *F. enigmaticus* exposed to 25 mg/L of MB for 24 h. * = $p < 0.05$; ** = $p < 0.01$; *** = $p < 0.001$.

	Reef	1 h	2 h	4 h	24 h
D%	Living	32.75 ± 8.57	43.71 ± 11.73	50.20 ± 12.11	80.43 ± 8.83
	Dead	15.06 ± 4.19	27.66 ± 5.61	33.49 ± 4.28	64.61 ± 5.94
	Difference	17.70 **	13.44 *	15.20 *	15.22 *
qt (mg/g)	Living	0.17 ± 0.04	0.21 ± 0.05	0.30 ± 0.08	0.53 ± 0.04
	Dead	0.08 ± 0.02	0.13 ± 0.02	0.15 ± 0.02	0.31 ± 0.03
	Difference	0.09 ***	0.08 **	0.15 **	0.22 **

Figure 2. Change in the color of the MB seawater solutions (top pair of photos) and in dead *Ficopomatus* reef blocks (bottom pair of photos) at the start of the experiment of exposure to MB solution (0 h) and at the end of the experiment (after 24 h).

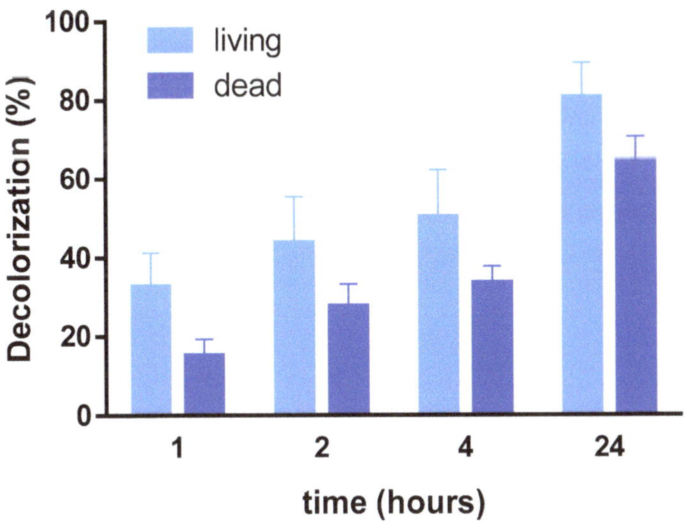

Figure 3. Percentage of decolorization (*D*%) of dead and living reefs over time.

3.2. *Reef Surface Characterization*

The observation of the surface of the dead and living reefs under a stereomicroscope highlighted a strong MB adsorption, as shown in Figure 4.

Figure 4. High-resolution details of tube surface after 24 h of exposure to MB seawater solution. Sample from dead reef.

The results of the ATR-FTIR analysis, necessary to identify the components likely involved in the adsorption process, are reported in Table 2, and the relative spectra are depicted in Figure 5. A high match (84.27%) was obtained with a reference composite material whose main components were found to be cellulose (contributing 60.52%), travertine (contributing 22.64%), and calcite (contributing 13.27%). The signal of MB was detectable and accounted for 3.57%.

Table 2. Percentage of matching of *F. enigmaticus* reef with reference composite materials and their relative percentage composition.

Total Match with Reference (%)	Name Composite	Composite Relative %
84.27	Cellulose	60.52
	Travertine	22.64
	Calcite	13.27
	Methylene blue, certified	3.57

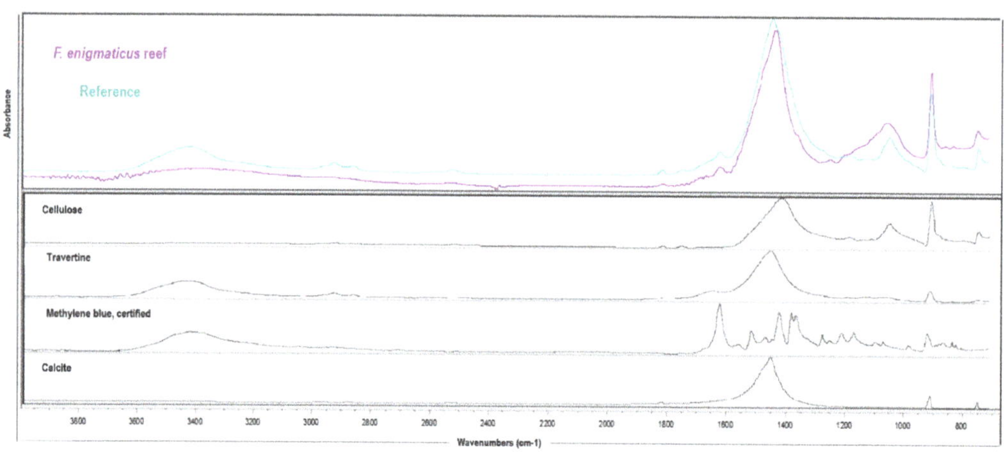

Figure 5. ATR-FTIR spectra of dead reefs of *F. enigmaticus* after exposure to 25 mg/L of methylene blue solution.

3.3. Toxicity Test

Higher toxicity values were observed in the non-bioremediated MB sample, resulting in a 59.8% inhibition of *A. fischeri* bioluminescence (Table 3). The solutions bioremediated by living reefs demonstrated toxicity levels ranging from 21.5% to 58.2%, with a dose-dependent response where increased residual concentrations of MB corresponded to greater toxicity. The samples bioremediated by dead reefs exhibited a toxic effect ranging from 30.5% to 38.2%. Seawater samples from the negative control (without MB but exposed to the reefs) displayed a significant ecotoxicological response, which was more pronounced in the case of the living reef (43.2%) than in the case of the dead reef (33.1%).

Table 3. Results of the toxicological test with *A. fischeri* (15 min, 90%) conducted on 24 h bioremediated solutions of MB using living (L) and dead (D) reefs. Values are listed in decreasing order of the tested residual MB concentrations. CP refers to the solution with non-bioremediated MB, while CN denotes seawater without MB exposed to living (CN_L) and dead (CN_D) reefs.

Samples	Residual MB Concentration Tested (mg/L)	*Aliivibrio fischeri* Bioluminescence Inhibition (%)	
		Mean	SD
CP	22.50	59.8	3.4
L2	8.39	58.2	0.0
L1	8.04	37.2	1.5
L5	6.09	38.5	0.0
L8	5.43	31.3	0.1
L7	2.99	31.1	0.0
L4	1.89	21.5	1.4
CN_L	0.00	43.4	1.5
D1	11.61	33.0	1.5
D2	9.58	30.5	1.9
D3	8.76	35.1	0.0
D5	8.04	31.3	0.7
D4	7.90	38.2	1.8
CN_D	0.00	33.1	1.9

4. Discussion

F. enigmaticus efficiently removed up to 81% of methylene blue (MB), a potential contaminant in aquatic systems resulting from industrial and domestic activities, within 24 h. Previous research has primarily focused on the removal of MB using biological systems, such as bacteria [28,33,34], fungi [35,36], marine algae [37–39], fruit [40,41], and snail shells [42], which were employed as biotechnology to treat dye-contaminated water after undergoing various processes like transformation, grinding, and encapsulation with nanomaterials. Less is known about the capacity of aquatic organisms to remove synthetic dyes once they are introduced into the environment, with limited information in the literature mostly focusing on plants, macroalgae, and microalgae. For instance, Manghabati et al. (2014) [43] reported a removal rate of 50–75% of MB (initial concentration: 40 mg/L) by the marine alga *Bifurcaria bifurcata* within 24 h. Al-Fawwaz et al. (2016) [44] observed a 30% removal of the dye by the green alga *Desmodesmus* sp. after 48 h of contact time. Santaeufemia et al. (2021) [29] investigated the percentage of dye removed by the marine algae *Phaeodactylum tricornutum* at different initial concentrations of MB and reported a 26.5% removal after 11 h (initial concentration of 20 mg/L). Also, Al-Baldawi et al. (2018) [45] studied the phyto-transformation of MB from water using the aquatic plant *Azolla pinnata* and reported a maximum degradation efficiency of 90% after 5 days at an initial concentration of 25 mg/L. The only study reporting performance similar to that obtained with *F. enigmaticus* in the present study is Imron et al. (2019) [26], where they achieved an 80.5% removal of MB within 24 h using the duckweed *Lemna minor* (initial concentration: 50 mg/L). Therefore, the efficiency of *F. enigmaticus* in removing the synthetic dye MB within just 24 h of contact time is comparable to, if not higher than, what has been obtained with other organisms, highlighting the high potential of this organism for bioremediation.

Besides the quantification of the capacity of *Ficopomatus* to remove MB, our findings also shed light on the processes underlying bioremediation. Two primary mechanisms have been hypothesized as the basis for the bioremediation capacity of *Ficopomatus*: bioaccumulation and bioadsorption. Evidence of the former can be qualitatively observed in microscopic images of the de-tubed worm body, which is distinctly colored blue after 24 h. This ability of MB to impart color is well-known and has contributed to its popularity in biological research, earning it the distinction as a vital dye par excellence (it stains negatively charged cellular components like nucleic acids). Notably, MB is recognized as one of the most valuable tools in various fields of biology, including histology, cytology, histochemistry, parasitology, mycology, microbiology, and virology, among others [46]. A comparison between the results obtained with the living and dead reefs, however, allowed us to estimate a limited contribution (18%) of bioaccumulation to decolorization after 24 h, supporting the hypothesis that the removal of MB could mostly depend on bioadsorption.

Bioadsorption was demonstrated through the microscopic and ATR-FTIR analysis of the reef surface, which facilitated the identification of the composition of the surface layer characterizing *Ficopomatus* tubes, thereby revealing the components likely involved in the process. The main component consisted of 60.52% cellulose followed by calcium carbonate in the form of calcite and travertine (altogether accounting for 35.91%). The MB signal was identified as further evidence of its adsorption. Therefore, the tubes predominantly featured cellulose as the main component, likely attributable to the layer of microalgae that develops over time on the external surface of the tube, imparting its characteristic brownish color [47]. Evidence of MB biosorption by microalgae is extensively documented in the literature. For example, Seoane et al. (2022) [38] used the biomass of the microalga *Chlamydomonas moewusii* for the removal of MB and obtained satisfactory results [38]; Yang et al. (2021) investigate the adsorption of MB on biochars derived from *Chlorella* sp. and *Spirulina* sp. [48], while Moghazy and Mahmoud (2023) used a macro-hollow loofah fiber, both with and without bio-attaching with the green microalga *Chlamydomonas reinhardtii*, for the removal of MB from an aqueous solution [49]. Such findings suggest that the hard structure of the reef itself, which was responsible for 61.51% of the removal of MB, could play a crucial role for contaminant sequestration, largely due to the presence of microalgae. Therefore, even when transient or persistent declines in *F. enigmaticus* population occur, such as after irregular river floods [50], their complex dead bioconstructions still ensure the continuity of much of the water remediation function, serving as collectors and catalysts for contaminant sequestration based on their associated microbiota.

No attempts have been made in this study to quantify the desorption of MB from *Ficopomatus* reefs, which requires further investigations to understand the potential of these organisms as dye bioremediators. Evidence from studies conducted on by-products of vegetal origin (therefore more similar to the surface composition of the *Ficopomatus* reefs of our study) suggest that the desorbing percentage of MB is generally limited and often requires the addition of solvents and several cycles of extraction. For instance, Daneshvar et al., (2017) [51] conducted a desorbing experiment with the brown macroalga *Nizamuddinia zanardinii*, which was used as a natural sorbent for MB removal from aqueous solutions. They achieved a desorption yield of only 2.36% using simple distilled water and had to prepare a mixture of 1 M HCl (25%)/1 M 1-butanol (75%) to enhance the performance to 64%. Similarly, Oladoja et al. (2009) [52] studied the batch desorption of MB from loofah using different solvents; NaOH and water yielded scarce results (<5.6% of MB desorbed), while acidic solvents (i.e., CH3COOH and HCl) were necessary to achieve 33% and 42%, respectively.

The acute toxicity tests on *Aliivibrio fischeri* revealed surprising and unexpected findings. Generally, the solutions proved to be toxic to *A. fischeri*, which is not surprising considering the well-documented antiseptic properties of MB in the literature [53]. However, bioremediated solutions from living reefs were found to be more toxic than those from dead reefs and, in relation to the residue, even more toxic than the positive control. In the first case, samples with similar residues from living (L1) and dead (D5) reefs caused a 37.2% and 31.3% inhibition of bioluminescence, respectively. In the second case, a similar response (58.2% vs. 59.8%) was triggered by 8.39 mg/L of bioremediated solutions and 22.50 mg/L of non-bioremediated solutions. Based on these data, it could be hypothesized that 8 mg/L may represent a threshold concentration beyond which the response curve of *A. fischeri* to MB no longer exhibits linear growth but instead becomes exponential. However, it is worth noting that the mere presence of *Ficopomatus* (CN) has proven to be harmful in itself (43.4% inhibition).

Two potential mechanisms of toxicity can be hypothesized to explain the observed pattern. The first is that MB may have been partially or completely metabolized into by-products, which are also toxic. MB is a basic phenothiazine dye that can undergo demethylation, resulting in the formation of azure A, azure B, azure C, benzidine and methylene [17,54]. For example, azure B has been shown to delay the development of the gastrula and blastula stages in the starfish *Marthasterias glacialis* [55]. However, which and how many substances of MB could be biotransformed still remain an essential gap in bioremediation experiments, requiring much more effort in future research. In this regard, techniques such as liquid chromatography (LC) and liquid chromatography coupled with mass spectrometry (LC/MS) can be capable of simultaneously determining a number of dyes, as reported in the review of Hakami and colleagues [56]. The second possible explanation is that *Ficopomatus* has the ability to produce compounds that interfere with quorum sensing (QS) in *A. fischeri*. QS is a synchronization mechanism within a bacterial population. Bacteria employ QS to communicate, regulate their behavior, assess their population density, and generate a synchronized behavior such as bioluminescence, virulence, or aggregation to form biofilms [57]. Here, we detected an effect of *Ficopomatus* reefs to inhibit the QS related to bioluminescence, but it is conceivable that the production of *Ficopomatus* secretions capable of disrupting bacterial QS is related to the formation of bacterial biofilms, suggesting a possible antifouling action. Despite evidence of the antifouling effects of metabolites from marine organisms, mostly focused on sponges (e.g., [58]), cnidarians (e.g., [59]), bryozoans (e.g., [60]), and ascidians (e.g., [61]), marine polychaetes are known to produce a wide range of antifouling, antipathogen, and defensive chemicals [62], reinforcing the hypothesis that *F. enigmaticus* could produce bioactive compounds. However, we may not exclude toxic effects on *A. fischeri* due to allelopathic compounds produced by the microbiota associated with reef structures (e.g., diatoms, [63]).

5. Conclusions

Our study on *F. enigmaticus* demonstrates its efficient removal of the synthetic dye methylene blue (MB) within just 24 h of contact time. Such findings unveil the potential role of this suspension-feeding and reef-building species, so far mostly considered as a threat to marine environments due to its invasiveness, to contribute to ecosystem services by enhancing water quality.

The slight difference in MB removal capacity between living and dead reefs (less than 18%) highlighted the ability of the calcified portion of reefs to sequester contaminants even after the colonies die. Moreover, dead reefs, by providing a suitable substrate, may represent natural collectors of microalgal biofilms that seemed to be involved in the process of adsorption of the dissolved dye. The high performance of *F. enigmaticus* in removing MB with respect to other organisms thus probably stems from the combination of direct removal by the worms and the indirect effects of reducing dye concentrations related to its nature as a habitat former.

The results of our preliminary assessment bode well for a possible employment of *F. enigmaticus* as a bioremediator, at least in specific environmental contexts (e.g., confined environments characterized by high levels of pollution) where potential negative effects of its presence are of relatively minor importance or the eradication of this species is not strictly necessary. Further investigations to extend the assessments to other classes of contaminants and to delve into the spatio-temporal dynamics underlying bioaccumulation are needed to fully explore the putative bioremediation abilities of *F. enigmaticus* and to help the management of this invasive species.

Supplementary Materials: The following supporting information can be downloaded at https://www.mdpi.com/article/10.3390/w16030368/s1: Table S1: Composition and physical-chemical properties of the synthetic seawater used for bioassay with *Aliivibrio fischeri*; Table S2: pH values measured at time 0 and after 24 h of contact time with 25 mg/L methylene blue. CN = negative controls; L = living reefs; D = dead reefs; CP = positive controls.

Author Contributions: Conceptualization, M.P., S.B., M.R. and V.V.; methodology, M.P., S.B. and M.R.; formal analysis, M.P.; investigation, M.P., V.V., S.A., T.B. and F.P.; resources, S.B. and M.R.; writing—original draft preparation, M.P.; writing—review and editing, S.B., M.P. and M.R. All authors have read and agreed to the published version of the manuscript.

Funding: This research received no external funding.

Data Availability Statement: The data presented in this study are available on request from the authors.

Acknowledgments: We express our gratitude to the Marine Biology Station Piran research group for its invaluable support in the Slovenian territory during the identification of *Ficopomatus* populations. Additionally, we appreciate their efforts in facilitating the transfer and exchange of information on the ecology and behavior of *Ficopomatus*, which proved to be a significant source of inspiration for this research.

Conflicts of Interest: The authors declare no conflicts of interest.

References

1. Dittmann, S.; Rolston, A.; Benger, S.N.; Kuptiyanova, E.K. Habitat Requirements, Distribution and Colonisation of the Tubeworm *Ficopomatus enigmaticus* in the Lower Lakes and Coorong. Report for the South Australian Murray-Darling Basin Natural Resources Management Board. 2009. Available online: https://www.researchgate.net/publication/260981354_Habitat_requirements_distribution_and_colonisation_of_the_tubeworm_Ficopomatus_enigmaticus_in_the_Lower_Lakes_and_Coorong (accessed on 1 December 2023).
2. Tebble, N. A source of danger to harbour structures—Encrustation by a tubed marine worm. *J. Instit. Munic. Engin.* **1953**, *80*, 259–265.
3. Fauvel, P. Un nouveau Serpuliden d'eau saumatre *Mercierella* n. g., *enigmatica* n. sp. *Bull. Soc. Zool. Fr.* **1923**, *47*, 424–430.
4. DAISIE. *Handbook of Alien Species in Europe*; Sringer: Dordrecht, The Netherlands, 2009; ISBN 978-1-4020-8279-5.
5. Charles, M.; Faillettaz, R.; Desroy, N.; Fournier, J.; Costil, K. Distribution, associated species and extent of biofouling "reefs" formed by the alien species *Ficopomatus enigmaticus* (Annelida, Polychaeta) in marinas. *Estuar. Coast. Shelf Sci.* **2018**, *212*, 164–175. [CrossRef]
6. Schwindt, E.; De Francesco, C.G.; Iribarne, O.O. Individual and reef growth of the invasive reef-building polychaete *Ficopomatus enigmaticus* in a south-western Atlantic coastal lagoon. *J. Mar. Biol. Ass. U.K.* **2004**, *84*, 987–993. [CrossRef]
7. Bruschetti, M.; Bazterrica, C.; Fanjul, E.; Luppi, T.; Iribarne, O. Effect of biodeposition of an invasive polychaete on organic matter content and productivity of the sediment in a coastal lagoon. *J. Sea Res.* **2011**, *66*, 20–28. [CrossRef]
8. Vaas, K.F. Immigrants among the animals of the delta-area of the SW. Netherlands. *Hydrobiol. Bull.* **1975**, *9*, 114–119. [CrossRef]
9. Bianchi, C.N.; Morri, C. *Ficopomatus* 'Reefs' in the Po River Delta (Northern Adriatic): Their Constructional Dynamics, Biology, and Influences on the Brackish-water Biota. *Mar. Ecol.* **1996**, *17*, 51–66. [CrossRef]
10. Bonifazi, A.; Lezzi, M.; Ventura, D.; Lisco, S.; Cardone, F.; Gravina, M.F. Macrofaunal biodiversity associated with different developmental phases of a threatened Mediterranean *Sabellaria alveolata* (Linnaeus, 1767) reef. *Mar. Environ. Res.* **2019**, *145*, 97–111. [CrossRef]
11. Heiman, K.W.; Micheli, F. Non-native Ecosystem Engineer Alters Estuarine Communities. *Integr. Comp. Biol.* **2010**, *50*, 226–236. [CrossRef]
12. Schwindt, E.; Iribarne, O.O. Reef of *Ficopomatus enigmaticus* (polychaeta; Serpulidae) in the Mar Chiquita Coastal Lagoon, Argentina. *Boll. Soco Hist. Nat. Balears* **1998**, *41*, 35–40.
13. Bruschetti, M.; Luppi, T.; Fankul, E.; Rosenthal, A.; Iribarne, O. Grazing effect of the invasive reef-forming polychaete *Ficopomatus enigmaticus* (Fauvel) on phytoplankton biomass in a SW Atlantic coastal lagoon. *J. Environ. Mar. Biol. Ecol.* **2008**, *354*, 212–219. [CrossRef]
14. Eno, N.C.; Clark, R.A.; Sanderson, W.G. *Non-Native Marine Species in British Waters: A Review and Directory*; JNCC: Peterborough, UK, 1997; ISBN 86107 442 5.

15. Oladoye, P.O.; Ajiboye, T.O.; Omotola, E.O.; Oyewola, O.J. Methylene blue dye: Toxicity and potential elimination technology from wastewater. *Results Eng.* **2022**, *16*, 100678. [CrossRef]
16. Bathia, D.; Sharma, N.R.; Singh, J.; Kanwar, R.S. Biological methods for textile dye removal from wastewater: A Review. *Crit. Rev. Environ. Sci. Technol.* **2017**, *47*, 1836–1876. [CrossRef]
17. Al-Tohamy, R.; Ali, S.S.; Li, F.; Okasha, K.M.; Mahmoud, Y.A.-G.; Elsamahy, T.; Jiao, H.; Fu, Y.; Sun, J. A critical review on the treatment of dye-containing wastewater: Ecotoxicological and health concerns of textile dyes and possible remediation approaches for environmental safety. *Ecotoxicol. Environ. Saf.* **2022**, *231*, 113160. [CrossRef] [PubMed]
18. Khan, I.; Saeed, K.; Zekker, I.; Zhang, B.; Hendi, A.H.; Ahmad, A.; Ahmad, S.; Zada, N.; Ahmad, H.; Ali, S.; et al. Review on Methylene Blue: Its Properties, Uses, Toxicity and Photodegradation. *Water* **2022**, *14*, 242. [CrossRef]
19. Al Farraj, D.A.; Elshikh, M.S.; Al Khulaifi, M.M.; Hadibarata, T.; Yuniarto, A.; Syafiuddin, A. Biotransformation and Detoxification of Antraquione Dye Green 3 using halophilic *Hortaea* sp. *Int. Biodeterior. Biodegrad.* **2019**, *140*, 72–77. [CrossRef]
20. Lu, G.; Nagbanshi, M.; Gouldau, N.; Jorge, M.M.; Meissner, P.; Jahn, A.; Mockenhaupt, F.P.; Muller, O. Efficacy and safety of methylene blue in the treatment of malaria: A systematic review. *BMC Med.* **2018**, *16*, 59. [CrossRef] [PubMed]
21. Pushparajah, M.; Renita, S.M.; Lielbelt, E.L. Methylene Blue an Antidote for Methemoglobinemia and Beyond. *Pediatr. Emerg. Care* **2021**, *37*, 474–477. [CrossRef]
22. Meng, X.; Li, Y.; Liu, Y.; Zhou, R.; Fu, Y.; Chen, J. Degradation of organic pollutants through activating bisulfite with lanthanum ferrite-loaded biomass carbon. *RSC Adv.* **2023**, *13*, 24819. [CrossRef]
23. Lipej, L.; Mavric, B.; Orlando-Bonaca, M.; Malej, A. State of the Art of the Marine Non-Indigenous Flora and Fauna in Slovenia. *Mediterr. Mar. Sci.* **2012**, *13*, 243–249. [CrossRef]
24. De Marchi, L.; Oliva, M.; Freitas, R.; Neto, V.; Figueira, E.; Chiellini, F.; Morelli, A.; Soares, A.M.V.M.; Pretti, C. Toxicity evaluation of carboxylated carbon nanotubes to the reef-forming tubeworm *Ficopomatus enigmaticus* (Fauvel, 1923). *Mar. Environ. Res.* **2019**, *143*, 1–9. [CrossRef] [PubMed]
25. Kazemi, F.; Mohamadnia, Z.; Kaboudin, B.; Karimi, Z. Photodegradation of methylene blue with a titanium dioxide/polyacrylamide photocatalyst under sunlight. *J. Appl. Polym. Sci.* **2016**, 43386. [CrossRef]
26. Imron, M.F.; Kurniawan, S.B.; Soegianto, A.; Wahudianto, F.E. Phytoremediation of methylene blue using duckweed (*Lemna minor*). *Heliyon* **2019**, *5*, e02206. [CrossRef] [PubMed]
27. Tan, K.A.; Morad, N.; Ooi, J.Q. Phytoremediation of Methylene Blue and Methyl Orange Using *Eichhornia crassipes*. *Int. J. Environ. Sci. Dev.* **2017**, *7*, 724–728. [CrossRef]
28. Contreras, M.; Grande-Tovar, C.D.; Vallejo, W.; Chaves-Lopez, C. Bio-Removal of Methylene Blue from Aqueous Solution by *Galactomyces geotrichum* KL20A. *Water* **2019**, *11*, 282. [CrossRef]
29. Santaeufemia, S.; Abalde, J.; Torres, E. Efficient removal of dyes from seawater using as biosorbent the dead and living biomass of the microalga *Phaeodactylum tricornutum*: Equilibrium and kinetics studies. *J. Appl. Phycol.* **2021**, *33*, 3071–3090. [CrossRef]
30. Toth, G.; Hahn, J.; Kriszt, B.; Szoboszlay, S. Acute and chronic toxicity of herbicides and their mixtures measured by *Aliivibrio fischeri* ecotoxicological assay. *Ecotoxicol. Environ. Saf.* **2019**, *185*, 109702. [CrossRef] [PubMed]
31. Dailianis, S.; Vlastos, D.; Zoppou, C.; Moschopoulou, A.; Antonopoulou, M. Different isoforms of parabens into marine environment: Biological effects on the bacterium *Aliivibrio fischeri* and the marine mussel *Mytilus galloprovincialis*. *Sci. Total Environ.* **2023**, *900*, 165902. [CrossRef]
32. *UNI EN ISO 11348-2:2019*; Qualità dell'acqua—Determinazione dell'effetto inibitorio di campioni acquosi sull'emissione di luce di Vibrio fischeri (prova su batteri luminescenti)—Parte 2: Metodo con batteri disidratati. Ente Nazionale Italiano di Unificazione (UNI): Milano, Italy, 2019.
33. Sarioglu, O.F.; Keskin, N.O.S.; Celebioglu, A.; Tekinay, T.; Uyar, T. Bacteria encapsulated electrospun nanofibrous webs for remediation of methylene blue dye in water. *Colloids Surf. B Biointerfaces* **2017**, *152*, 245–251. [CrossRef]
34. Deb, M.; Redkar, N.; Manohar, C.S.; Jagtap, A.S.; Saxena, S.; Shukla, S. *Bacillus* sp. based nano-bio hybrids for efficient water remediation. *Environ. Pollut.* **2023**, *326*, 121490. [CrossRef]
35. Liu, J.; Su, H.; Xue, J.; Wei, X. Optimization of Decoloration Conditions of Methylene Blue Wastewater by *Penicillium* P1. *Indian. J. Microbiol.* **2022**, *62*, 103–111. [CrossRef] [PubMed]
36. Lino, V.; Castaldo, R.; Gentile, G.; Manini, P. Reusable melanin-based biosorbents for efficient methylene blue removal: The new frontier of fungi-inspired allomelanin coatings for sustainable water remediation processes. *Mater. Today Sustain.* **2023**, *21*, 100283. [CrossRef]
37. Bouzikri, S.; Ouasfi, N.; Benzidia, N.; Salhi, A.; Bakkas, S.; Khamliche, L. Marine alga "*Bifurcaria bifurcata*": Biosorption of Reactive Blue 19 and methylene blue from aqueous solutions. *Environ. Sci. Pollut. Res.* **2020**, *27*, 33636–33648. [CrossRef]
38. Seoane, R.; Santaeufemia, S.; Abalde, J.; Torres, E. Efficient Removal of Methylene Blue Using Living Biomass of the Microalga *Chlamydomonas moewusii*: Kinetics and Equilibrium Studies. *Int. J. Environ. Res. Public. Health* **2022**, *19*, 2653. [CrossRef] [PubMed]
39. Pathak, V.V.; Kothari, R.; Chopra, A.K.; Singh, D.P. Experimental and kinetic studies for phycoremediation and dye removal by *Chlorella pyrenoidosa* from textile wastewater. *J. Environ. Manag.* **2015**, *163*, 270–277. [CrossRef] [PubMed]
40. Mehmandost, N.; Goudarzi, N.; Chamjangali, M.A.; Bagherian, G. Removal of methylene blue and crystal violet in binary aqueous solution by magnetic *Terminalia catappa* kernel shell biosorbent using Box–Behnken design. *J. Iran. Chem. Soc.* **2022**, *19*, 3769–3781. [CrossRef]

41. Saha, P. Assessment on the Removal of Methylene Blue Dye using Tamarind Fruit Shell as Biosorbent. *Water Air Soil. Pollut.* **2010**, *213*, 287–299. [CrossRef]
42. Sharma, A.; Devi, I. A sustainable biosorption technique for treatment of industrial wastewater using snail shell dust (*Bellamya bengalensis*). *Environ. Monit. Assess.* **2023**, *195*, 389. [CrossRef]
43. Manghabati, H.; Pazuki, G. A study on the decolorization of methylene blue by *Spirodela polyrrhiza*: Experimentation and modeling. *RSC Adv.* **2014**, *4*, 30137. [CrossRef]
44. Al-Fawwaz, A.; Adbullah, M. Decolorization of Methylene Blue and Malachite Green by Immobilized *Desmodesmus* sp. Isolated from North Jordan. *Int. J. Environ. Sci. Dev.* **2016**, *7*, 95–99. [CrossRef]
45. Al-Baldawi, I.A.; Abdullah, S.R.S.; Anuar, N.; Hasan, H.A. Phytotransformation of methylene blue from water using aquatic plant (*Azolla pinnata*). *Environ. Technol. Innov.* **2018**, *11*, 15–22. [CrossRef]
46. Barbosa, P.; Peters, T.M. The effects of vital dyes on living organisms with special reference to Methylene Blue and Neutral Red. *Histochem. J.* **1971**, *3*, 71–93. [CrossRef] [PubMed]
47. Ten Hove, H.A.; Weerdenburg, C.A. A generic revision of the brackish-water serpulid *Ficopomatus* southern 1921 (polychaeta: Serpulinae), including *Mercierella* fauvel 1923, *Sphaeropomatus* treadwell 1934, mercierellopsis rioja 1945 and *Neopomatus* pillai 1960. *Biol. Bull.* **1978**, *154*, 96–120. [CrossRef] [PubMed]
48. Yang, Z.; Hou, J.; Miao, L.; Wu, J. Comparison of adsorption behavior studies of methylene blue by microalga residue and its biochars produced at different pyrolytic temperatures. *Environ. Sci. Pollut. Res.* **2021**, *28*, 14028–14040. [CrossRef] [PubMed]
49. Moghazy, R.M.; Mahmoud, R.H. Microalgal-based macro-hollow loofah fiber bio-composite for methylene blue removal: A promising step for a green adsorbent. *Int. J. Biol. Macromol.* **2023**, *253*, 127009. [CrossRef] [PubMed]
50. Aliani, S.; Bianchi, C.N.; Meloni, R. Scanning electron microscope observations on the tube of the reef-forming serpulid *Ficopomatus enigmaticus* (Fauvel) (Annelida, Polychaeta). *Boll. Zool.* **1995**, *62*, 363–367. [CrossRef]
51. Daneshvar, E.; Vazirzadeh, A.; Niazi, A.; Kousha, M.; Naushad, M.; Bhatnagar, A. Desorption of Methylene blue dye from brown macroalga: Effects of operating parameters, isotherm study and kinetic modeling. *J. Clean. Prod.* **2017**, *152*, 443–453. [CrossRef]
52. Oladoja, N.A.; Aboluwoye, C.O.; Akinkugbe, A.O. Evaluation of Loofah as a Sorbent in the Decolorization of Basic Dye Contaminated Aqueous System. *Ind. Eng. Chem. Res.* **2009**, *48*, 2786–2794. [CrossRef]
53. Kacem, I.; Laurent, T.; Blanchemain, N.; Neut, C.; Chai, F.; Haulton, S.; Hildebrand, H.F.; Martel, B. Dyeing and antibacterial activation with methylene blue of a cyclodextrin modified polyester vascular graft. *J. Biomed. Mater. Res. Part A* **2014**, *102A*, 2942–2951. [CrossRef]
54. Zhang, X.; Hui, Y.; Fang, C.; Wang, Y.; Han, F.; Lou, X.; Fodjo, E.K.; Cai, Y.; Kong, C. Determination of Methylene Blue and Its Metabolite Residues in Aquatic Products by High-Performance Liquid Chromatography–Tandem Mass Spectrometry. *Molecules* **2021**, *26*, 4975. [CrossRef]
55. Brice, M. Toxicité des colorants metachromatiques pour l'oeuf d'etoil de mer. *Archs Biol.* **1959**, *70*, 121–125.
56. Hakami, A.A.H.; Wabaidur, S.M.; Khan, M.A.; AlOthman, Z.A.; Siddiqui, M.R. Extraction Procedures and Analytical Methods for the Determination of Methylene Blue, Rhodamine B and Crystal Violet—An Overview. *Curr. Anal. Chem.* **2021**, *17*, 708–728. [CrossRef]
57. Miller, M.B.; Bassler, B.L. QUORUM SENSING IN BACTERIA. *Annu. Rev. Microbiol.* **2001**, *55*, 165–199. [CrossRef] [PubMed]
58. Tintiller, F.; Moriou, C.; Petek, S.; Fauchon, M.; Hellio, C.; Saulnier, D.; Ekins, M.; Hooper, J.N.A.; Al-Mourabit, A.; Debitus, C. Quorum Sensing Inhibitory and Antifouling Activities of New Bromotyrosine Metabolites from the Polynesian Sponge *Pseudoceratina* n. sp. *Mar. Drugs* **2020**, *18*, 272. [CrossRef] [PubMed]
59. Zhang, J.; Liang, Y.; Liao, X.-J.; Deng, Z.; Xu, S.-H. Isolation of a new butenolide from the South China Sea gorgonian coral *Subergorgia suberosa*. *Nat. Prod. Res. Former. Nat. Prod. Lett.* **2014**, *28*, 150–155. [CrossRef] [PubMed]
60. Kon-ya, K.; Shimidzu, N.; Adachi, K.; Miki, W. 2,5,6-Tribromo-l-methylgramine, an Antifouling Substance from the Marine Bryozoan *Zoobotryon pellucidum*. *Fish. Sci.* **1994**, *60*, 773–775. [CrossRef]
61. Olguin-Uribe, G.; Abou-Mansour, E.; Boulander, A.; Debard, H.; Francisco, C.; Combaut, G. 6-Bromoindole-3-carbaldehyde, from an *Acinetobacter* sp. bacterium associated with the ascidian *Stomozoa murrayi*. *J. Chem. Ecol.* **1997**, *23*, 2507. [CrossRef]
62. Coutinho, M.C.L.; Teixeira, V.L.; Santos, C.S.G. A Review of "Polychaeta" Chemicals and their Possible Ecological Role. *J. Chem. Ecol.* **2018**, *44*, 72–94. [CrossRef]
63. Ianora, A.; Miralto, A. Toxigenic effects of diatoms on grazers, phytoplankton and other microbes: A review. *Ecotoxicology* **2010**, *19*, 493–511. [CrossRef]

water

MDPI

Article

ZnO-CuO Nanocomposite as an Efficient Adsorbent for As(III) Removal from Water

Jesús Plácido Medina Salas [1,*], Francisco Gamarra Gómez [1], Elisban Juani Sacari Sacari [1,2,*], Wilson Orlando Lanchipa Ramos [1], Rocío María Tamayo Calderón [3], Efracio Mamani Flores [4], Víctor Yapuchura Platero [5], Walter Dimas Florez Ponce de León [6] and Elmer Marcial Limache Sandoval [7]

1 Laboratorio de Nanotecnología (NanoLab), Facultad de Ingeniería, Universidad Nacional Jorge Basadre Grohmann, Av. Miraflores s/n, Tacna 23003, Peru; fgamarrag@unjbg.edu.pe (F.G.G.)
2 Centro de Energías Renovables de Tacna (CERT), Facultad de Ciencias, Universidad Nacional Jorge Basadre Grohmann, Av. Miraflores s/n, Tacna 23003, Peru
3 Centro de Microscopia Electrónica, Facultad de Ingenieria de Procesos, Universidad Nacional de San Agustín, Arequipa 04001, Peru
4 Departamento de Física Aplicada, Facultad de Ciencias, Universidad Nacional Jorge Basadre Grohmann, Av. Miraflores s/n, Tacna 23003, Peru
5 Departamento de Ingeniería Civil, Facultad de Ingeniería Civil, Arquitectura y Geotecnia, Universidad Nacional Jorge Basadre Grohmann, Av. Miraflores s/n, Tacna 23003, Peru
6 Departamento de Química, Facultad de Ingeniería, Universidad Nacional Jorge Basadre Grohmann, Av. Miraflores s/n, Tacna 23003, Peru
7 Grupo de Investigación HIDROCIENCIA, Facultad de Ciencias de la Salud, Universidad Privada de Tacna, Av. Jorge Basadre Grohmann s/n, Pocollay, Tacna 23000, Peru
* Correspondence: jmedinas@unjbg.edu.pe (J.P.M.S.); esacaris@unjbg.edu.pe (E.J.S.S.)

Abstract: Arsenic (III) exposure, often from contaminated water, can have severe health repercussions. Chronic exposure to this toxic compound is linked to increased risks of various health issues. Various technologies exist for arsenic (III) removal from contaminated water sources. This work synthesized ZnO-CuO nanocomposites through ultrasound-assisted coprecipitation, generating abundant hydroxylated sites via the deposition of ZnO nanoparticles onto CuO sheets for enhanced arsenic (III) adsorption. Structural characterization verified the formation of phase-pure heterostructures with emergent properties. Batch studies demonstrated exceptional 85.63% As(III) removal at pH 5, where binding with prevalent neutral H_3AsO_3 occurred through inner-sphere complexation with protonated groups. However, competing anions decreased removal through site blocking. Favorable pseudo-second order chemisorption kinetics and the 64.77 mg/g maximum Langmuir capacity revealed rapid multilayer uptake, enabled by intrinsic synergies upon nanoscale mixing of Zn/Cu oxides. The straightforward, energy-efficient ultrasonic production route makes this material promising for real-world water treatment integration.

Keywords: nanocomposite; ZnO-CuO; arsenic; water treatment; ultrasonication; chemisorption

1. Introduction

Contamination of water by toxic heavy metals is a major issue that has serious consequences for human health and the natural environment [1]. While heavy metals occur naturally, human activities like mining, industrial processes, and agricultural runoff contribute to their release into the environment [2]. Consequently, heavy metals accumulate in water, soil, and air, posing a threat to plants, animals, and humans [3].

One such heavy metal is arsenic, which exists naturally but is also discharged into the environment through human practices like mining, smelting, and coal burning [4,5].

High arsenic levels have been reported in groundwater and tap water in regions of Asia, including Bangladesh, India, and China [6,7]. Alarming concentrations are also prevalent in some Latin American countries, such as Argentina [8], Chile [9], Mexico [10], and Peru [11,12]. In Peru specifically, arsenic contamination is linked to mining activities and the abundance of volcanic systems with high arsenic content [12,13]. The city of Tacna, located in southern Peru, has reported arsenic concentrations of up to 0.5 mg/L in drinking water from reservoirs recharged by arsenic-rich volcanic rocks [14]. This level far exceeds the 10 μg/L limit recommended by the World Health Organization [15].

Arsenic contamination of water sources presents serious health risks to humans. Exposure to arsenic can result in various health problems [16], including skin lesions [17], gastrointestinal and neurological issues [18,19], cancer [20,21], as well as birth defects and developmental challenges in children [22].

Arsenic (III) is the predominant form of arsenic in water and is notably toxic compared to other forms [23], making its removal challenging. Arsenic (III) acts as a reducing agent and can chemically transform when exposed to oxygen, resulting in its conversion to arsenic (V), which exhibits reduced toxicity and improved water removal capabilities [24].

A broad spectrum of techniques and technologies can be utilized for arsenic removal from water, but their efficacy varies based on factors such as arsenic type, concentration, and cost [25]. Common technologies include precipitation, coagulation, adsorption, and ion exchange [26]. Precipitation employs chemicals to induce arsenic precipitation, while coagulation facilitates the clumping of arsenic particles for easier removal [27]. Adsorption employs materials to bind arsenic to their surfaces, and ion exchange uses resins to exchange arsenic ions for other ions [28].

Adsorption is the process by which a substance adheres to the surface of another substance and offers significant advantages over other technologies. It is highly efficient, cost-effective, versatile, and safe for arsenic removal from water [29]. For this purpose, different materials that could remove arsenic from water have been studied, such as TiO_2 [30] and their composites [31,32], iron-based materials [33,34], nickel-based materials [35], perovskite-type materials, such as $CaTiO_3$ [36], graphene-based materials [37], carbon nanotubes [38], and other materials [39]. In terms of effectiveness and cost, the adsorption method is the most efficient, but its operation is still expensive.

CuO, ZnO, and ZnO/CuO nanomaterials are promising adsorbents for arsenic remediation [40–43] on account of their substantial surface area and porous properties [44], hydroxyl functional groups on the surface [45], variable surface charges [46], redox activity [47], wide pH stability [48], reusability [49], low cost, and easy synthesis methods [50–56]. These properties facilitate arsenic adsorption through surface complexation, ion exchange, electrostatic interactions, coprecipitation, and redox reactions. However, the specific mechanisms of arsenite adsorption are not well understood and require further research to optimize nanosorbent design for efficient arsenic removal [57].

The main objectives of this research involved the synthesis of ZnO/CuO nanocomposite at different concentrations using an ultrasound-assisted precipitation method. Subsequently, the synthesized materials were subjected to comprehensive characterization to evaluate the impact of the composition on their structural, vibrational, and morphological properties, as well as their effectiveness in arsenic (III) removal. In this investigation, a comprehensive set of analytical methods was utilized. These techniques included X-ray diffraction (XRD), Fourier-transform infrared (FTIR) spectroscopy, scanning electron microscopy (SEM), and BET surface area. Furthermore, batch experiments to evaluate the arsenic adsorption capabilities of the synthesized materials were also performed.

2. Materials and Methods

2.1. Materials

All the chemicals employed in this investigation were of analytical quality and were utilized without any additional purification steps. 2-Propanol, Zinc (II) Acetate Dihydrate, Copper (II) Nitrate Trihydrate, Sodium Hydroxide, Hydrochloric Acid (37%), Sodium (meta) Arsenite, Sodium Chloride, Sodium Nitrate, and Sodium Sulfate were supplied by Merck Company (Darmstadt, Germany).

2.2. ZnO-CuO Nanocomposites Synthesis

The synthesis of the ZnO-CuO nanocomposite was achieved by employing the ultrasound-assisted precipitation method. For this purpose, five samples were synthesized with 00:100 (0: 3.9922 g), 25:75 (1.09745 g: 2.99475 g), 50:50 (2.195 g: 1.9965 g), 75:25 (3.2925 g: 0.99825 g), and 100:00 (4.3923 g: 0) molar ratios of zinc (II) acetate dihydrate and copper (II) nitrate trihydrate, respectively. Each of these reagents was mixed and dissolved by a 35 kHz ultrasound cleaner (VWR Symphony, Radnor, PA, USA) in 500 mL of 2-propanol for 30 min and labeled as an A solution. Similarly, 0.08 M (1.6 g) of sodium hydroxide was dissolved in 500 mL of ultrapure water for 30 min and labeled as a B solution. After 30 min of sonication, the B solution was added dropwise to the A solution and then kept in constant sonication for an additional 30 min. Subsequently, the resulting solid was subjected to several rounds of ultrapure water washing, followed by filtration, drying at room temperature, and finally, calcination at 400 °C for a period of 2 h. The obtained samples were labeled CuO, ZnO-CuO (25:75), ZnO-CuO (50:50), ZnO-CuO (75:25), and ZnO, according to the molar ratio.

2.3. Characterization

The X-ray diffraction (XRD) patterns of the ZnO-CuO nanocomposites were recorded using a diffractometer from AERIS Research (Malvern Panalytical Ltd., Almedo, The Netherlands) with Ni-filtered CuKα radiation (wavelength 1.5406 Å) over a 2θ range from 20° to 80°. The crystallite size and structural parameters were determined by Rietveld refinement using X'Pert HighScore Plus software version 4.9. Nitrogen adsorption analysis at 77 K was carried out on a Gemini VII surface area analyzer from Micromeritics Instrument Co. (Norcross, GA, USA) to measure the specific surface area (BET) of the nanocomposites after degassing the sample at 200 °C under helium flow for 2 h. Attenuated total reflectance Fourier-transform infrared (ATR-FTIR) spectroscopy over the 400–4000 cm^{-1} range was performed using a Bruker Invenio R spectrometer (Ettlingen, Germany) to acquire the vibrational spectra. A Thermo Scientific Quattro S field emission scanning electron microscope (FE-SEM) (Thermo Scientific Co., Eindhoven, The Netherlands) equipped with an UltraDry EDS detector was used to analyze the morphology of the samples and perform elemental analysis, and Thermo Scientific Talos F200i Transmission Electron Microscope (Thermo Scientific Co., Eindhoven, The Netherlands) was utilized to examine the shape and particle size of the samples. The arsenic (III) concentration was quantified using a flameless atomic absorption spectrometer (Shimadzu AA-6300, Shimadzu Scientific Instruments, Inc., Kyoto, Japan) with a graphite furnace atomizer (GFA-EX7i), employing a hollow cathode lamp (293.7 nm, 25 mA) for arsenic and a deuterium lamp for background correction.

2.4. Batch Adsorption Experiments

A 10 ppm arsenic (III) stock solution was prepared by dissolving 5.45 mg of sodium meta-arsenite ($NaAsO_2$) in 1 L of ultrapure water to provide the arsenic (III) adsorbate.

Batch adsorption experiments were then performed to evaluate the arsenic removal capacity of the synthesized ZnO-CuO nanocomposite materials. These batch tests served as a rapid screening tool for analyzing the adsorption capabilities of the nanocomposites by studying the depletion in arsenic(III) concentration when equilibrated with the nanocomposite adsorbent material under controlled conditions. In the experimental procedure, a solution consisting of 100 mL of water containing 10 ppm of arsenic (III) at pH 7 was subjected to a treatment involving the combination of 50 mg/L of adsorbent. The resultant solution was then stirred at a rate of 200 rpm for 2 h while maintaining a temperature of 25 degrees Celsius. The arsenic adsorption experiments were conducted in a darkened environment. Samples were collected at various time intervals; then, centrifugation was carried out at 6000 rpm for 5 min, and thereafter, the supernatant was analyzed utilizing graphite furnace atomic absorption spectroscopy.

2.5. Impact of pH on Arsenic Removal

To assess the impact of pH on the effectiveness of arsenic removal, a series of solutions was prepared. These solutions maintained a constant concentration of 10 ppm of arsenic (III) while varying the pH levels. The pH was changed by adding solutions of either sodium hydroxide or diluted hydrochloric acid. After an adsorption procedure that lasted for two hours, samples were collected, and subsequently, they were subjected to centrifugation at 6000 revolutions per minute. A graphite furnace atomic absorption spectrometer was used to analyze the supernatant.

The equilibrium adsorption isotherms of the sample with the better performance were studied by conducting batch experiments under dark conditions using 50 mg of the ZnO-CuO nanocomposite and varying initial arsenic (III) concentrations of 10, 20, 40, 60, 80, and 100 ppm dissolved in 100 mL of deionized water. The resultant adsorption data after 2 h of contact time were modeled using the linearized Langmuir and Freundlich isotherm equations.

The Langmuir model assumes monolayer coverage of energetically equivalent sites, while the Freundlich model accounts for multilayer adsorption on a heterogeneous surface. Analysis of the correlation coefficients and chi-square values from the linear regression fitting revealed that the Langmuir model better described the arsenic (III) adsorption behavior.

2.6. PZC of ZnO-CuO Nanocomposite

The point of zero charge (PZC) of ZnO-CuO nanocomposites with the most exceptional adsorption capabilities was determined using the salt addition method [58]. To elucidate, 40 mL of a 0.1 M $NaNO_3$ solution was aliquoted into nine separate Erlenmeyer flasks. The pH levels of the ZnO-CuO nanocomposite suspensions, spanning the range from 3 to 11, were adjusted by introducing either 0.1 M nitric acid or 0.1 M sodium hydroxide solutions, monitored with a pH meter (HI 5221, HANNA Instruments, Woonsocket, RI, USA). Following this, 0.02 g of ZnO-CuO nanocomposites were introduced into each flask and agitated at 150 rpm on an orbital shaker, maintaining a temperature of 30 °C for a duration of 24 h. Once equilibrium was reached, the contents were filtered, and the pH values of the filtrates were duly recorded. Subsequently, the PZC value, representing the pH at which the total surface charge on the adsorbent becomes neutral, was determined by plotting a graph correlating the initial pH with the change in pH.

2.7. Influence of Coexisting Ions on Arsenic Removal

To examine the effects of competing ions, test solutions were made using sodium salts of chloride, nitrate, and sulfate at varying concentrations of 100, 250, and 500 ppm.

The adsorption experiments were performed with an initial arsenic (III) concentration of 10 ppm and a dosage of 50 mg/L of adsorbent. A sodium hydroxide solution was employed to adjust the pH of the solutions to ~7.0.

After the batch adsorption tests were conducted for two hours, samples were collected from the test solutions containing the ZnO-CuO nanocomposite adsorbent and arsenic species. These samples were centrifuged at 4000 rpm for 5 min to separate out the adsorbent particles. Centrifugation facilitated the isolation of the supernatant from the solid adsorbent phase. The supernatant consisted of the residual arsenic species remaining in the solution after the adsorption process. This supernatant was then analyzed using a sensitive atomic absorption spectroscopy technique coupled with a graphite furnace atomizer.

3. Results and Discussion

3.1. X-ray Diffraction Results

X-ray diffraction (XRD) analysis provided critical insights into the crystalline phase composition, nanostructure, and structural parameters of the synthesized ZnO-CuO nanocomposites (Figure 1a). The samples contained two distinct components, hexagonal zincite ZnO (JCPDS 01-075-1526) and monoclinic tenorite CuO (JCPDS 00-005-0661), without any additional impurity peaks. The relative intensity of the ZnO and CuO peaks varied proportionally with their composition ratios, as expected.

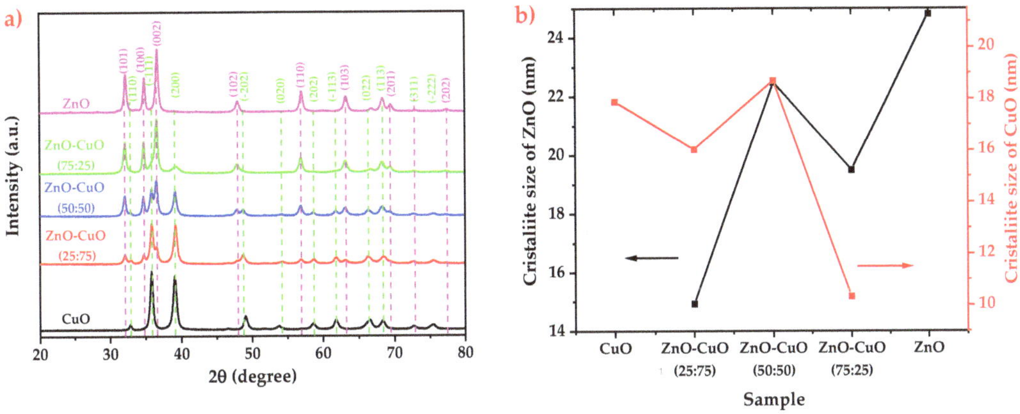

Figure 1. (**a**) X-ray diffraction profiles of ZnO-CuO nanocomposites and (**b**) Crystallite size of ZnO and CuO.

Detailed Rietveld refinement of the XRD patterns was performed using PANalytical HighScore Plus software V.4.9 to quantify the relative phase abundances, crystallite sizes, and lattice constants in the nanocomposites. The background was modeled using the Chebyshev function, and the pseudo-Voigt profile was applied for peak fitting. Table 1 summarizes the results of the nanocomposites.

Table 1. Structure-related parameters for ZnO-CuO nanocomposites.

Sample	CuO (100)	ZnO-CuO (25:75)		ZnO-CuO (50:50)		ZnO-CuO (75:25)		ZnO (100)
Phase	**CuO**	**ZnO**	**CuO**	**ZnO**	**CuO**	**ZnO**	**CuO**	**ZnO**
Crystal system	Monoclinic	Hexagonal	Monoclinic	Hexagonal	Monoclinic	Hexagonal	Monoclinic	Hexagonal
Phase percentage (%)	100	24.71	75.29	48.57	51.43	75.2	24.8	100
a (nm)	4.67793	3.25339	4.72394	3.25317	4.71857	3.25237	4.76458	3.24885
b (nm)	3.42191	3.25339	3.40375	3.25317	3.40595	3.25237	3.3736	3.24885
c (nm)	5.12759	5.21114	5.1312	5.21158	5.13222	5.21043	5.1187	5.20628
α (°)	90	90	90	90	90	90	90	90
β (°)	99.49249	90	99.95483	90	99.88692	90	99.96167	90
γ (°)	90	120	90	120	90	120	90	120
ρ (g/cm^3)	6.5	5.66	6.5	5.66	6.5	5.66	6.52	5.67
D (nm)	17.9	14.93	16.03	22.48	18.7	19.49	10.33	24.81
R_{exp} (%)	1.37993	1.32982		1.31176		1.32851		1.24266
R_{wp} (%)	2.23386	2.26934		2.24639		2.5136		2.16768
R_p (%)	1.74381	1.86191		1.96468		1.7781		1.58302
GOF	1.29957	1.31047		1.34613		1.25582		1.74439
SSA_{BET} (m^2/g)	22.3098 $\pm$ 0.0625 m^2/g	21.8229 $\pm$ 0.0198 m^2/g		17.7192 $\pm$ 0.2239 m^2/g		19.5901 $\pm$ 0.1120 m^2/g		18.4638 $\pm$ 0.0642 m^2/g

Excellent fitting confirmed that the samples contained highly crystalline zincite ZnO and tenorite CuO as the only phases with tunable relative compositions, as expected from the synthesis ratios. The ZnO crystallite size exhibited an initial increase from 14.93 nm to 22.48 nm as CuO content increased to 50%, which was attributed to CuO-mediated nucleation and growth. However, at 75% CuO, the size reduced to 10.33 nm, indicating dominant ZnO growth inhibition effects, as depicted in Figure 1b and outlined in Table 1. These trends imply that complex synergistic interactions between ZnO and CuO modulate crystallite growth during nanocomposite formation. Furthermore, the specific surface area understandably correlated inversely with crystallite size, as displayed in Table 1.

3.2. FTIR Spectral Analysis

Fourier-transform infrared spectroscopy was employed for the detailed characterization of the chemical environments and bonding structures in the synthesized ZnO-CuO nanocomposites (Figure 2). The FTIR spectra provided complementary evidence, alongside XRD and microscopy, on the successful formation of ZnO and CuO phases with tunable compositions.

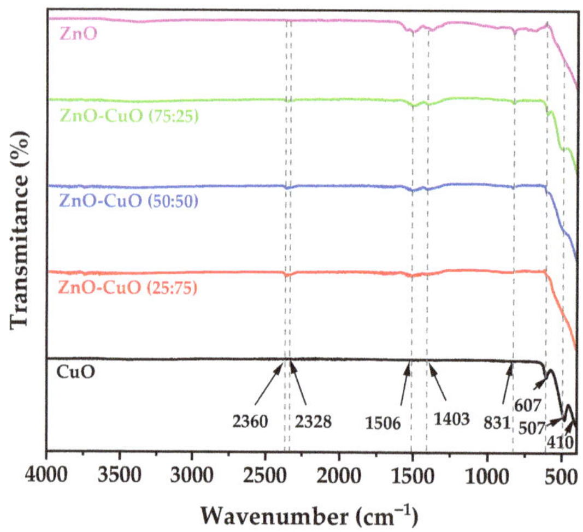

Figure 2. FTIR spectrums of ZnO-CuO nanocomposites.

The Zn-O and Cu-O metal–oxygen vibrational stretches confirmed the incorporation of ZnO and CuO, with relative intensities varying proportionally to the Zn and Cu content, as expected from the synthesis ratios. Specifically, the Zn-O peaks at 410 cm^{-1} and 607 cm^{-1} corresponded to the hexagonal wurtzite structure of the ZnO [59,60]. The Cu-O stretch at 507 cm^{-1} indicated the presence of a monoclinic CuO [61].

The FTIR spectra also exhibited small peaks at 2328 cm^{-1} and 2360 cm^{-1}, corresponding to the stretching vibrations of residual carboxyl groups. These leftover organic carboxyl moieties likely originated from the incomplete decomposition of the zinc acetate and 2-propanol precursors during the thermal treatment [62].

Additional signals observed at 1403 cm^{-1} and 1506 cm^{-1} were assigned to the symmetric stretching vibrations of carboxylate functionalities (COO^-) associated with residual acetate groups [63]. The relative intensities of the carboxyl and carboxylate peaks showed a direct correlation with the ZnO:CuO composition ratios. The higher the copper content, the greater the intensity of these peaks was observed. This variational trend aligns with the greater stability of copper acetate complexes compared to zinc acetates, leading to higher residuals at higher CuO contents.

The presence of residual organic groups like carboxylates and hydroxides revealed incomplete decomposition of the zinc acetate and 2-propanol precursors during the low-temperature ultrasonication treatment. This was likely responsible for the residual carbon observed in the EDS analysis (Table 2).

3.3. Scanning and Transmission Electron Microscopy

Field scanning electron microscopy (FE-SEM) and energy-dispersive X-ray spectroscopy (EDS) were applied to investigate both the morphology and elemental composition of ZnO-CuO composites. Figure 3 shows that the ZnO-CuO composites consist of CuO leaves covered by irregularly shaped ZnO nanoparticles (Figure 3). The increase in ZnO content tends to cover the CuO leaves progressively, as can be observed in Figure 3d,f,h), where CuO leaves were widely covered. EDS analysis was effectively utilized to validate and quantify the elemental composition. This technique allowed for the precise determination of the elemental ratios in ZnO-CuO composite materials. Table 2 reveals the relative abundance of Zn and Cu in the samples based on the elemental composition determined from the samples. Notably, the Zn/Cu ratio determined by EDS analysis closely matched both the anticipated values based on the molar composition used in the synthesis and the XRD results. This indicates that ZnO and CuO have been extensively incorporated, and the composites are homogeneous. Carbon content in samples is due to calcination residual carbon and organic volatilization from precursors.

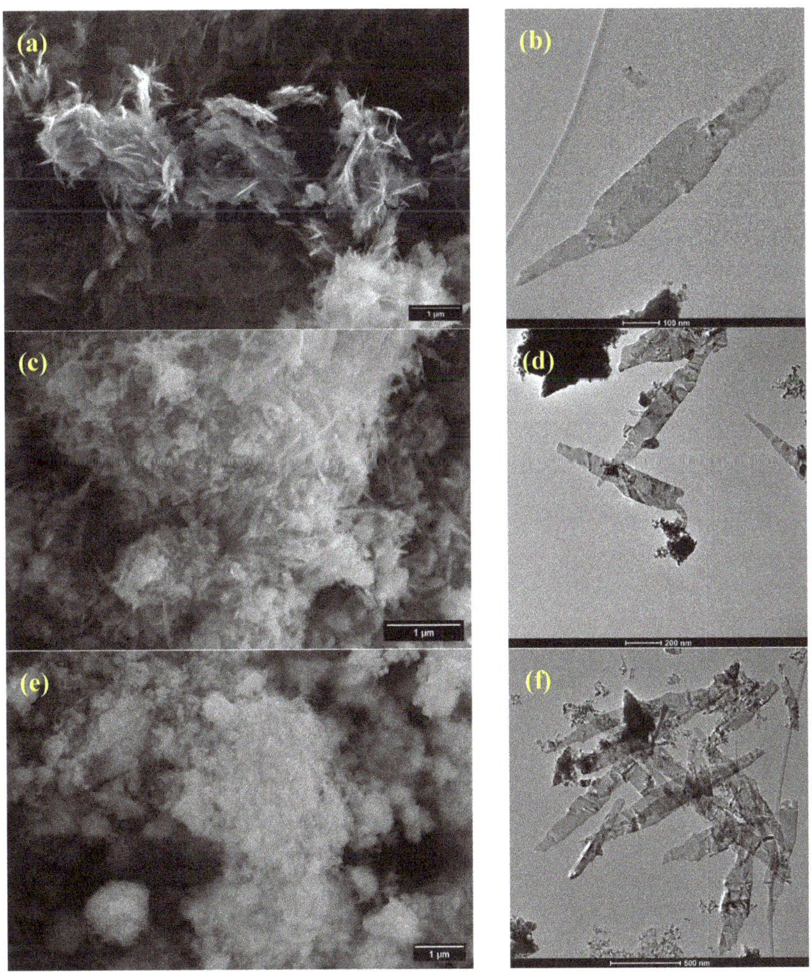

Figure 3. *Cont.*

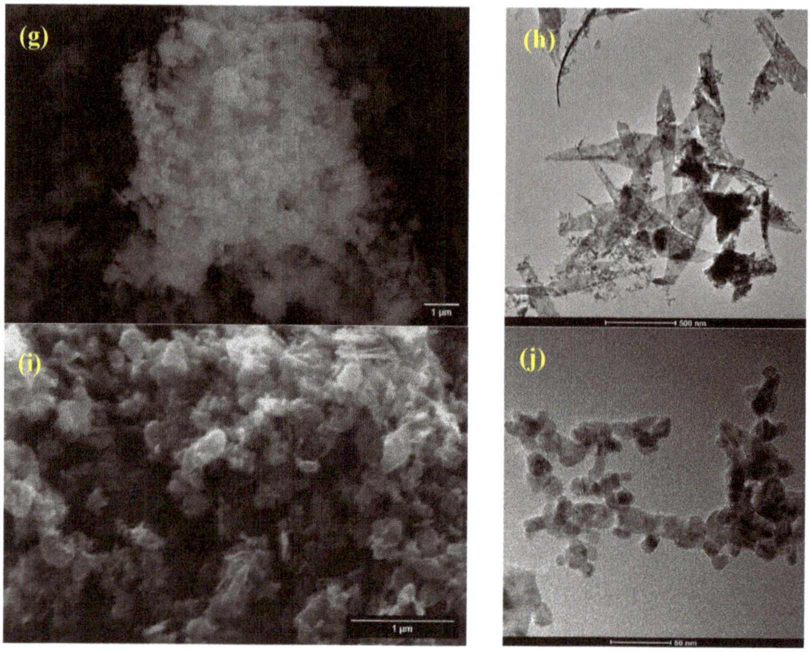

Figure 3. Scanning and Transmission Electron Microscopy images of (**a**,**b**) CuO, (**c**,**d**) ZnO-CuO (25:75), (**e**,**f**) ZnO-CuO (50:50), (**g**,**h**) ZnO-CuO (75:25), and (**i**,**j**) ZnO.

Transmission Electron Microscopy (TEM) images are displayed in Figure 3b,d,f,h,j), and the particle size distribution is displayed in Figure S2. Comparing the average sizes and standard deviations obtained after analyzing several images, the data reveal a clear trend. Pure CuO exhibited 477 ± 275 nm sized leaves, decreasing to 439 ± 260 nm in the ZnO-CuO (25:75) composite as small ZnO nanoparticles began adsorbing onto surfaces. However, the ZnO-CuO (50:50) composite showed considerably larger 702 ± 307 nm particles, implying a large surface for ZnO growth and generating severe aggregation from ZnO bridging between CuO leaves. Thereafter, the 75:25 ratio limited amassing with 487 ± 234 nm sizes. Pure ZnO nanoparticles possessed small 25 ± 7 nm irregular morphologies. Hence, the nanoparticulate decoration of ZnO onto CuO lever and pore surfaces seems to first assist crystalline growth before later inducing aggregation through interlinkage.

Table 2. EDS Elemental Analysis of ZnO-CuO Composite Materials.

Element	Sample				
	ZnO (100)	**ZnO-CuO (25:75)**	**ZnO-CuO (50:50)**	**ZnO-CuO (75:25)**	**CuO (100)**
Zn (Atom %)	30.42	5.85	15.93	19.95	-
Cu (Atom %)	-	20.33	17.00	6.07	27.46
O (Atom %)	19.79	60.24	57.04	59.69	61.32
C (Atom %)	49.79	13.58	10.03	14.29	11.21

3.4. Analysis of Arsenic (III) Adsorption

The arsenic (III) removal efficiency of the synthesized ZnO-CuO nanocomposites was evaluated in batch adsorption trials performed under pH 7 conditions, starting with an initial arsenic concentration of 10 mg/L (Figure 4). Prior to these experiments, meticulous calibration of the atomic absorption spectrometer was conducted, yielding an impressive

R-squared value of 0.999 (Figure S1). Then, three nanocomposite samples with varying ZnO:CuO molar ratios of 25:75, 50:50, and 75:25 were tested. It was observed that arsenic (III) removal efficiency was strongly dependent on the composition, with the ZnO-CuO (50:50) nanocomposite demonstrating a maximum removal of 70.5%. The ZnO-CuO (25:75) and ZnO-CuO (75:25) composites showed relatively lower removal rates of 43.02% and 63%, respectively. In comparison, pure CuO and ZnO nanoparticles exhibited relatively low removal rates of 1.02% and 0.41%, respectively. The enhanced performance of the 50:50 nanocomposite can be attributed to optimal synergistic effects between ZnO and CuO at this ratio for providing abundant hydroxylated surface binding sites for arsenic (III) adsorption [44,64]. The abundance of these hydroxyl groups facilitated arsenic (III) removal at pH 7 through mechanisms like surface complexation and ion exchange [46,65].

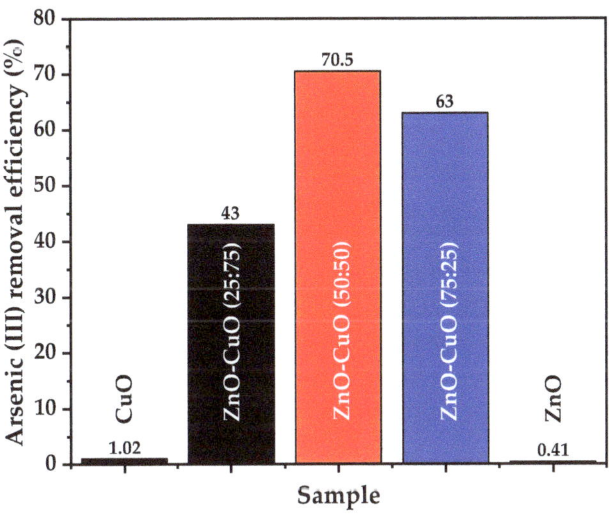

Figure 4. Arsenic removal efficiency of ZnO-CuO nanocomposite.

3.5. pH Influence on Arsenic (III) Removal

The influence of contact time on the arsenic (III) adsorption capacity of the 50:50 ZnO-CuO nanocomposite was evaluated at varying pH levels, as depicted in Figure 5a. It was evident that the amount of arsenic adsorbed rapidly increased during the initial stages of contact between the adsorbate and the adsorbent. This was followed by the attainment of adsorption equilibrium after 120 min for all tested pH conditions. The initial rapid uptake indicates fast diffusion and accessibility of the arsenic species to the binding sites on the nanocomposite surface. The plateau after 120 min signifies saturation of the available adsorption sites. The fast adsorption kinetics are beneficial for potential applications, as they would require shorter residence times.

Batch adsorption experiments revealed a maximum arsenic (III) removal of 85% by the ZnO-CuO (50:50) nanocomposite at pH 5 after 2 h. The abundant hydroxyl groups on the nanocomposite surface, coupled with the positive surface charge under acidic conditions, promoted favorable chemisorption interactions with arsenic (III) species. Speciation studies show that at pH 5, arsenic exists predominantly as neutral H_3AsO_3, which can directly bind to positively charged surface sites via inner-sphere complexation.

The kinetics of arsenic (III) adsorption by the ZnO-CuO nanocomposites under varying pH conditions were analyzed using two common kinetic models—the pseudo-first order and pseudo-second order rate equations [36].

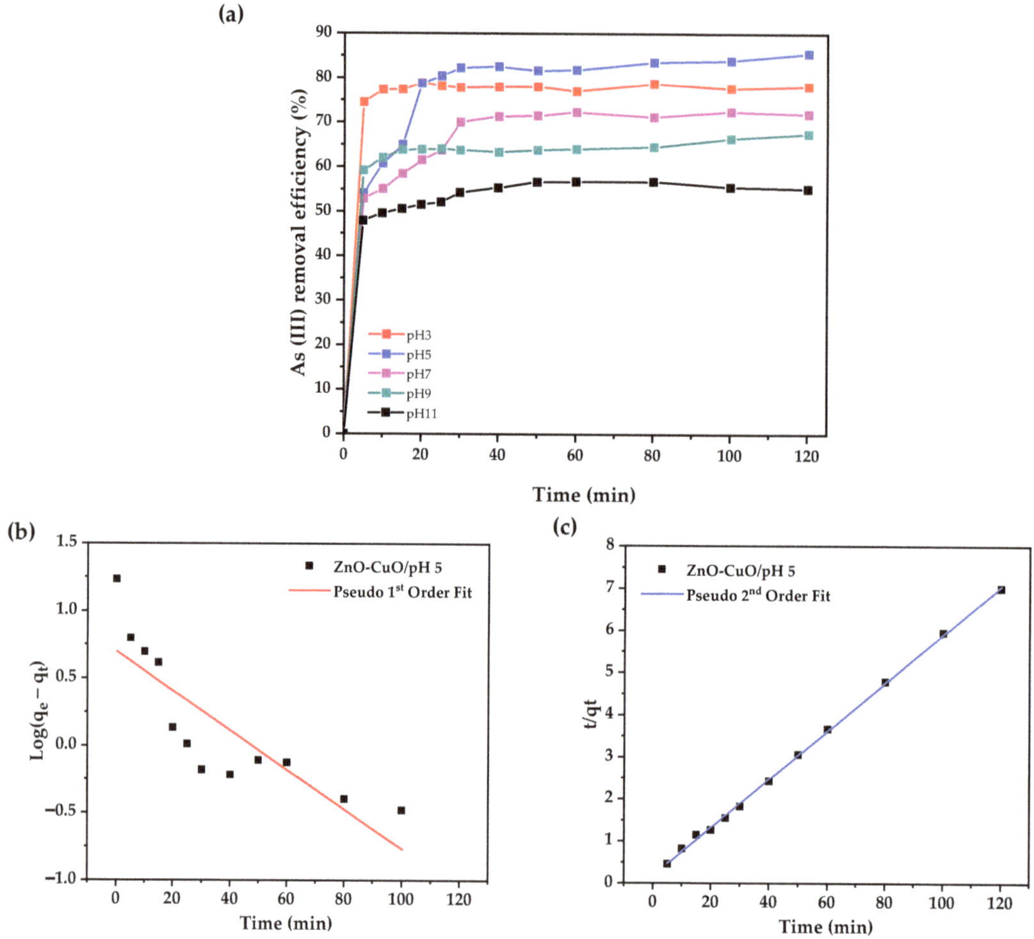

Figure 5. (**a**) Time-dependent arsenic (III) adsorption by ZnO-CuO (50:50) at various pH, with the initial arsenic concentration set at 10 mg/L and adsorbent concentration at 0.5 g/L; (**b**) Linear fitting of the kinetic adsorption utilizing the pseudo-first-order model; and (**c**) Linear fitting using the pseudo-second-order model.

The pseudo-first-order model (Equation (1)) assumes that adsorption occurs through a physisorption mechanism with weak interactions between the adsorbate and the adsorbent surface. The pseudo-second-order model (Equation (2)) assumes that chemisorption is the rate-limiting step involving stronger chemical bonding via the exchange or sharing of electrons between adsorbent and adsorbate [66].

$$\frac{dq_t}{dt} = k_1(q_e - q_t) \quad (1)$$

$$\frac{dq_t}{dt} = k_2(q_e - q_t)^2 \quad (2)$$

where q_t denotes the adsorption capacity at time t; q_e denotes the adsorption capacity at equilibrium, while k_1 and k_2 represent the rate constants for pseudo-first and pseudo-second-order kinetics, respectively. Equations (3) and (4) provide the linearized forms of these kinetic expressions.

$$log\frac{(q_e - q_t)}{q_e} = -\frac{k_1}{2.303}t \quad (3)$$

$$\frac{t}{q_t} = \frac{1}{k_2 q_e^2} + \frac{t}{q_e} \quad (4)$$

The obtained kinetic coefficients are presented in Table 2.

The results show that the kinetic data fitted both models (Figure 5b,c) and present decent correlation coefficients (R^2), as shown in Table 3, indicating that the arsenite adsorption follows multiple steps [67,68]. However, the pseudo-second-order model demonstrated a better fit overall, with R^2 values very close to 1. According to Ho et al. [67], the excellent fit of the experimental kinetic data to the pseudo-second-order model suggests that chemisorption is likely to be the rate-limiting step controlling the kinetics of arsenic adsorption by the ZnO-CuO nanocomposite.

Table 3. Kinetic coefficients from pseudo-first-order and pseudo-second-order fitting models based on ZnO-CuO nanocomposites experiment data.

pH	Pseudo 1st Order			Pseudo 2nd Order		
	k_1	q_e	R^2	k_2	q_e	R^2
pH 3	0.01369084	1.19	0.60288	0.62686683	15.63	0.99991
pH 5	0.00634824	2.01	0.70912	0.01764271	17.48	0.99918
pH 7	0.02101172	3.36	0.85998	0.02300911	14.88	0.99959
pH 9	0.01300043	2.30	0.54333	0.04975609	13.44	0.99909
pH 11	0.02208858	2.18	0.98732	0.08071776	11.29	0.99925

Chemisorption involves the formation of strong chemical bonds between the adsorbate and adsorbent surface atoms. This typically occurs through the sharing or exchanging of valence electrons, creating coordinated covalent linkages [69].

Analysis of the pseudo-second-order rate constants (k_2) revealed that the k_2 values progressively decreased as the solution pH increased from acidic to alkaline conditions.

The higher k_2 and faster adsorption kinetics under acidic pH can be attributed to the positive surface charge of the nanocomposite and favorable electrostatic interactions with the negatively charged arsenic (III) ions. Under acidic conditions, the hydroxylated ZnO-CuO surface becomes protonated and acquires a net positive charge [43]. This enhances electrostatic attraction between the positively charged surface binding sites and the anionic arsenite species ($H_3AsO_3^-/HAsO_3^{2-}$), accelerating the rate of chemisorption [70,71]. However, as the pH increases, deprotonation of the surface hydroxyl groups occurs, imparting a negative charge to the nanocomposite surface. This electrostatic repulsion between the now negatively charged surface and anionic arsenite ions likely reduces the chemisorption rate [24,72].

Linear forms of Langmuir and Freundlich models were utilized to characterize the equilibrium adsorption data (Table 4).

Table 4. Adsorption isotherms models and their linear forms.

Isotherm Model	Nonlinear Form	Linear Form	Plot	References
Langmuir	$q_e = \frac{Q_L K_L C_e}{1+K_L C_e}$	$\frac{1}{q_e} = \frac{1}{K_L q_{max}}\frac{1}{C_e} + \frac{1}{q_{max}}$	$\frac{1}{q_e}$ vs. $\frac{1}{C_e}$	[73]
Freundlich	$q_e = K_F C_e^{1/n}$	$lnq_e = lnK_f + \frac{1}{n}lnC_e$	lnq_e vs. lnC_e	[74]

The Langmuir equation assumes monolayer coverage with uniform binding sites, allowing the calculation of the maximum uptake capacity (q_{max}). Its linear form permits the determination of the dimensional equilibrium constant K_L, signifying adsorbate-adsorbent affinity. Conversely, the empirical Freundlich model presumes multilayer sorption across heterogeneous localities of variant affinities. Hence, binding escalates indefinitely with concentration. The heterogeneity factor $1/n$ provides insights into sorption intensity.

By fitting experimental results to these fundamental isotherms and deriving key parameters, including equilibrium concentration *Ce*, quantity bound at equilibrium q_e, and constants that reveal critical binding tendencies, the adsorption mechanism, capacity, and favorability can be effectively elucidated from a theoretical lens to support the empirical observations.

To enhance comprehension of the adsorption process and ascertain the maximal capacity of arsenic (III) removal using ZnO-CuO (50:50) nanocomposite, adsorption isotherm data were acquired under ambient conditions at a temperature of 25 degrees Celsius. The data were subjected to fitting using the linear equations of the Langmuir and Freundlich models, respectively.

The results displayed in Figure 6 and Table 5 show a superior Freundlich model linear regression fit (R^2 = 0.9856) that verifies heterogeneous multilayer binding of arsenic (III) ions onto the variable ZnO-CuO (50:50) nanocomposite surface [74].

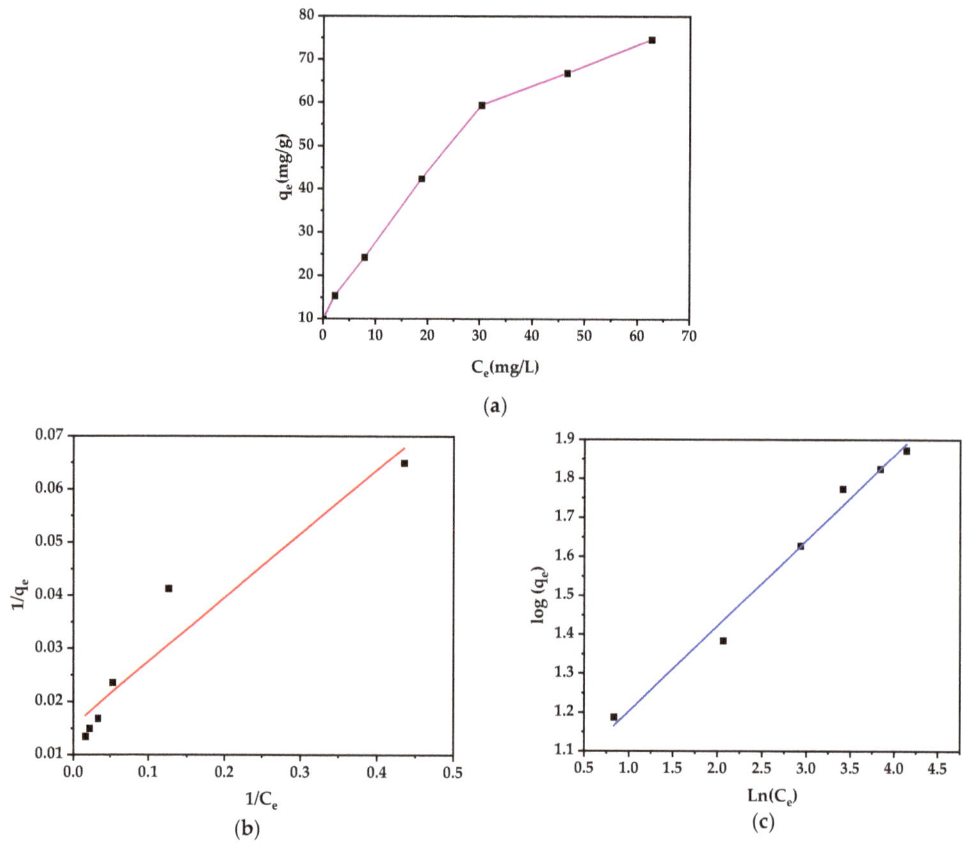

Figure 6. (**a**) Effect of initial concentration of Arsenic (III) removal by ZnO-CuO (50:50) nanocomposite; (**b**) Langmuir's isotherm plot; and (**c**) Freundlich's isotherm plots.

Table 5. Isotherm parameters for the arsenic (III) by ZnO-CuO.

Isotherm Model	Langmuir				Freundlich		
Parameters	q_{max} (mg/g)	K_L	R_L	R^2	K_f	$1/n$	R^2
Value	64.7668	0.1283	0.4380	0.9243	9.6168	0.2190	0.9856

At acidic pH 5, the neutral H_3AsO_3 species predominates and can access both high- and low-energy binding sites through a combination of inner-sphere direct surface complexation on abundant hydroxyls [75], as well as secondary interactions. This multilayer arrangement leads to a very high overall adsorption capacity [76].

The decent Langmuir fit suggests that while some monolayer coverage occurs at stronger chemisorption sites [77], the diverse surface enables extensive multilayer arsenic (III) adsorption through the spectrum of weaker physisorption sites [78].

The kinetics showed that chemisorption likely limits the initial uptake rate [79]. Thereafter, rapid diffusion and physisorption drive the high equilibrium capacities by allowing neutral H_3AsO_3 to saturate the heterogeneous surface of the nanocomposite [80].

Table 6 illustrates a comparison of several materials based on ZnO and CuO that have been reported for the removal of arsenic (As).

Through direct ultrasound-mediated nanostructuring, our ZnO-CuO (50:50) nanocomposite demonstrates an appreciable 64.77 mg/g arsenic (III) Langmuir adsorption capacity. This surpasses conventionally fabricated precursors, including pure ZnO (5.03 mg/g) [81] and ZnO-GO (8.17 mg/g) composites [82], validating the intrinsic benefits of synergistic nanotexturing. Notably, only tailor-made supports like solution combustion-derived ZnO-CuO/g-C_3N_4 [46] or dopants with citrate-hydrothermally synthesized Pd@ZnO/CuO [83] enable higher capacities near 100 mg/g. However, despite the absence of expensive adjuncts, our facile sonochemical synthesis elicits emergent morphological properties from interfacial Zn/Cu alloying, conferring substantial As(III) coordination strength.

Additionally, employing only non-toxic, cost-effective metal salts lends promise for sustainable, large-scale adoption. This establishes an easily prepared, competitively performing nano-adsorbent via component manipulation. Further stoichiometric modulation could elucidate mechanisms and systematic property enhancements.

Table 6. Comparative Arsenic Adsorption Capacities in ZnO-CuO Nanocomposites and Other Adsorbents.

Material	Synthesis Method	Test Conditions	q_{max} (mg/g)	Reference
ZnO-CuO/g-C_3N_4	Solution combustion method	7 mg of adsorbent; 50 mL of 20–150 ppb As (III) solution; 70 min	97.56	[46]
CuO-ZnO	Electrospinning	5 mg of adsorbent; 10 mL of 1–9 ppm As (III) solution; 24 h	26.27	[44]
ZnO-GO	Solvothermal	0.3 g of adsorbent; 100 mL of 10–50 ppm As (III) solution; 1 h.	8.17	[82]
CuO-ZnO doped chitosan succinic acid	Freeze-drying process	10 mg of adsorbent; 5 mL of 0.1–1.2 ppm As (III) solution; 24 h.	0.899	[84]
ZnO	Precipitation	10 mg of adsorbent; 4 mL of 3–1000 ppm As (III) solution	5.03	[81]
ZnO nanorods	Precipitation	0.4 g of adsorbent; 50 mL of 30–90 ppb As (III) solution	38.46	[85]
Pd@ZnO/CuO	Citrate-hydrothermal	10 mg of adsorbent; 10 mL of 250 ppb arsenic solution; 3 h	91.96	[83]
CuO/ZnO	Ultrasound	50 mg of adsorbent; 100 mL of arsenic solution with 10–100 ppm; 2 h	64.77	present work

3.6. PZC Analysis of ZnO-CuO Nanocomposite

The point of zero charge results are displayed in Figure 7. In contrast with the previous results, the maximum 85.63% arsenic (III) removal observed at pH 5 aligns closely with the point of zero charge (PZC) of 6.45 when the net surface charge is minimized, enabling favorable chemisorption [41]. Below the PZC, the positively charged surface attracted anionics $H_2AsO^{3-}/HAsO_3^{2-}$ through electrostatics, conferring 78.18% (pH 3) to 85.63% (pH 5) uptake. Above the PZC, surface repulsion of these species reduced capacity to 72.55% (pH 7), 67.55% (pH 9), and 55.57% (pH 11) as hydroxylated areas grew [86].

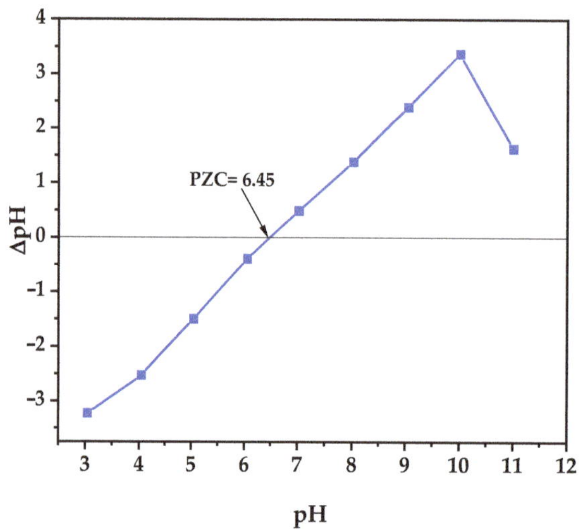

Figure 7. PZC plot for ZnO-CuO nanocomposite by salt addition method.

Explicitly integrating adsorption capacities across the pH range with PZC-governed surface charging provides in-depth insights into the multifaceted electrostatic and chemical factors influencing nano-adsorption [87]. The direct relationship of solution chemistry with binding verifies the intricate interplay of site competition, surface speciation, and charge repulsion in directing pH-responsive arsenic (III) removal.

3.7. Impact of Coexisting Ions

The adsorption of arsenic (III) by ZnO-CuO (50:50) nanocomposites at pH 7 occurs primarily through surface complexation mechanisms involving inner-sphere and outer-sphere binding to surface metal ions and hydroxyl groups [88,89]. The presence of competing anions can interfere with arsenic (III) adsorption by occupying these chemisorption sites or altering surface charge. As seen in Figure 8, increasing concentrations of chloride, nitrate, and sulfate ions (100 to 500 mg/L) at pH 7 all progressively reduced arsenic (III) removal efficiency by the nanocomposite. At the highest concentration (500 mg/L), chloride lowered removal from 72.9% to 47%, nitrate to 47.8%, and sulfate to 49.5%. The monovalent chloride ion likely competes directly with arsenite oxyanions for surface binding [83]. The multivalent sulfate and nitrate ions have greater hydration energies, displacing arsenic (III) through the outer-sphere complex formation [90,91]. Also, specific adsorption of sulfate and nitrate imparted a more negative surface charge at pH 7, weakening the arsenic (III) electrostatic attraction [92].

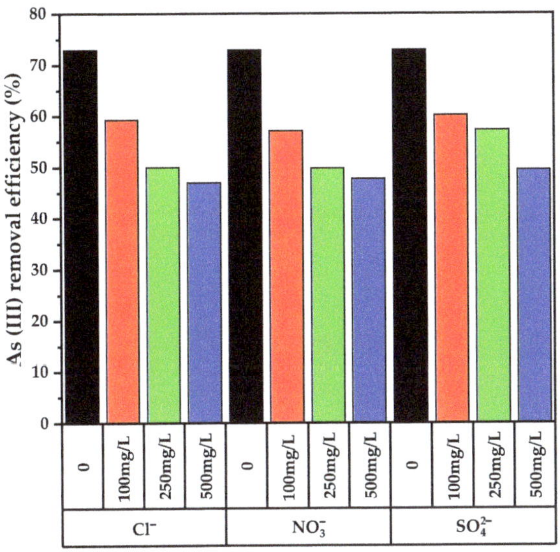

Figure 8. Influence of different competing ions on arsenic (III) removal using ZnO-CuO (50:50).

Despite current experimental limitations for statistical testing, the substantial arsenic adsorption capacity of the synthesized ZnO-CuO nanocomposite through sonochemical methods is evident in this evaluation.

3.8. Stability

The effectiveness of the ZnO-CuO (50:50) composite in removing arsenic (III) was the primary topic of an in-depth study that was conducted on the phase stability of the material under varied pH conditions. For the purpose of determining the structural stability of the composite material, an XRD study was carried out. The XRD patterns that were obtained by exposing the ZnO-CuO (50:50) composite to various pH conditions are displayed in Figure 9.

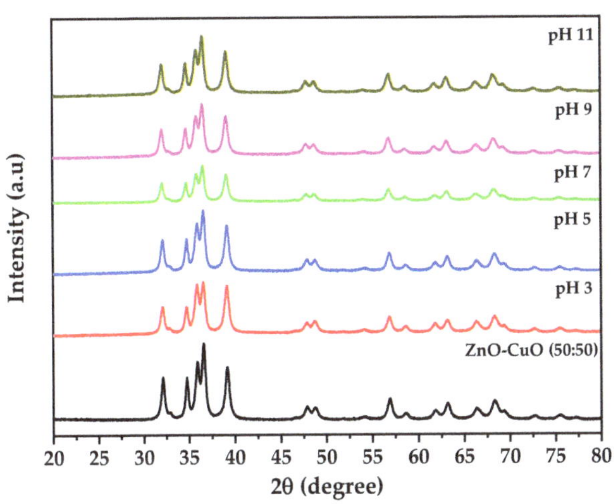

Figure 9. XRD patterns of ZnO-CuO (50:50) after different pH evaluations.

XRD results show that the ZnO-CuO (50:50) composite keeps its original structure across the whole pH range that was tested. The absence of any new phases and the absence of a significant decrease in the abundance of any phase served as evidence that the composite material has chemical stability under varied pH conditions.

The evaluation of arsenic (III) removal uncovered intriguing trends related to pH, which is important to note. At a pH of 5, the ZnO-CuO (50:50) composite showed the best arsenic removal effectiveness, whereas at a pH of 11, the composite showed the lowest removal efficiency. These findings are consistent with the XRD patterns that were found, providing more evidence that phase stability is directly related to arsenic removal efficiency.

4. Conclusions

The ultrasound-assisted coprecipitation method successfully produced optimized ZnO-CuO nanocomposites with a 50:50 molar ratio, demonstrating excellent 85.63% arsenic (III) elimination capacity under acidic conditions (pH 5) where chemisorptive binding with neutral H_3AsO_3 prevailed. Favorable pseudo-second order adsorption kinetics revealed rapid uptake, reaching equilibrium within 2 h. The nanomaterial maintained stability across a broad pH spectrum, retaining selectivity despite competing ions.

Multilayer Freundlich model dominance verified specialized heterogeneous surface textures that enable extensive physisorption-driven stacking, complementing chemisorption. The 64.77 mg/g maximum binding capacity, accessible ultrasonication synthesis, and intrinsic synergies upon nanoscale mixing of Zn/Cu oxides substantiate the promise for sustainable large-scale adoption to mitigate critical arsenic contamination issues.

Moreover, competing ions had a few significant effects on the arsenic adsorption by the ZnO-CuO (50:50) composite. Similarly, the stability of the composite was investigated, revealing that the ZnO-CuO (50:50) nanocomposite maintains its structure over a broad pH range. Thus, such a ZnO-CuO composite has true potential as a nanocomposite for water remediation applications.

Supplementary Materials: The following supporting information can be downloaded at: https://www.mdpi.com/article/10.3390/w15244318/s1, Figure S1. Graphite furnace Atomic absorption spectrometer calibration curve of arsenic (III). Figure S2. Particle size distribution of pure ZnO, CuO and ZnO-CuO nanocomposite.

Author Contributions: J.P.M.S. and F.G.G., funding acquisition, experiment design, investigation, and review and editing; E.J.S.S., experiment design, investigation, characterization, data analysis, and writing—original draft, review, and editing; W.O.L.R., investigation and characterization; R.M.T.C., E.M.F., V.Y.P. and W.D.F.P.d.L. performed data analysis and writing—original draft; E.M.L.S., investigation and characterization. All authors have read and agreed to the published version of the manuscript.

Funding: This research was funded by the Universidad Nacional Jorge Basadre Grohmann through "Fondos del canon sobrecanon y regalias mineras", approved by rectoral resolution N° 3780-2014-UN/JBG with the project "Study of the application of nanotechnology for the purification of water with arsenic in the Tacna region".

Data Availability Statement: The data presented in this study are available on request from the corresponding author.

Acknowledgments: We are grateful to the Universidad Nacional Jorge Basadre Grohmann and the projects "Development of thin films of MWCNT/TiO_2 to improve the efficiency of dye-sensitized solar cells (DSSC)" and "Determination of the optical fingerprints of solid, liquid and organic materials using visible and infrared spectroscopy", of the Universidad Nacional Jorge Basadre Grohmann, approved by rectoral resolutions No. 4516-2018-UN/JBG and No. 5854-2019-UN/JBG, respectively, for their facilities for materials characterization.

Conflicts of Interest: The authors declare no conflict of interest.

References

1. Sankhla, M.S.; Kumari, M.; Nandan, M.; Kumar, R.; Agrawal, P. Heavy Metals Contamination in Water and Their Hazardous Effect on Human Health—A Review. *Int. J. Curr. Microbiol. App. Sci.* **2016**, *5*, 759–766. [CrossRef]
2. Vareda, J.P.; Valente, A.J.M.; Durães, L. Assessment of heavy metal pollution from anthropogenic activities and remediation strategies: A review. *J. Environ. Manag.* **2019**, *246*, 101–118. [CrossRef]
3. Tchounwou, P.B.; Yedjou, C.G.; Patlolla, A.K.; Sutton, D.J. Heavy metal toxicity and the environment. *Exp. Suppl.* **2012**, *101*, 133–164. [CrossRef] [PubMed]
4. Garelick, H.; Jones, H.; Dybowska, A.; Valsami-Jones, E. Arsenic pollution sources. *Rev. Environ. Contam. Toxicol.* **2008**, *197*, 17–60. [CrossRef] [PubMed]
5. Han, F.X.; Su, Y.; Monts, D.L.; Plodinec, M.J.; Banin, A.; Triplett, G.E. Assessment of global industrial-age anthropogenic arsenic contamination. *Naturwissenschaften* **2003**, *90*, 395–401. [CrossRef] [PubMed]
6. Rahman, M.M.; Chowdhury, U.K.; Mukherjee, S.C.; Mondal, B.K.; Paul, K.; Lodh, D.; Biswas, B.K.; Chanda, C.R.; Basu, G.K.; Saha, K.C.; et al. Chronic arsenic toxicity in Bangladesh and West Bengal, India—A review and commentary. *J. Toxicol. Clin. Toxicol.* **2001**, *39*, 683–700. [CrossRef] [PubMed]
7. Chowdhury, U.K.; Biswas, B.K.; Chowdhury, T.R.; Samanta, G.; Mandal, B.K.; Basu, G.C.; Chanda, C.R.; Lodh, D.; Saha, K.C.; Mukherjee, S.K.; et al. Groundwater arsenic contamination in Bangladesh and West Bengal, India. *Environ. Health Perspect.* **2000**, *108*, 393–397. [CrossRef]
8. Litter, M.I.; Ingallinella, A.M.; Olmos, V.; Savio, M.; Difeo, G.; Botto, L.; Farfán Torres, E.M.; Taylor, S.; Frangie, S.; Herkovits, J.; et al. Arsenic in Argentina: Occurrence, human health, legislation and determination. *Sci. Total Environ.* **2019**, *676*, 756–766. [CrossRef]
9. Alam, M.A.; Mukherjee, A.; Bhattacharya, P.; Bundschuh, J. An appraisal of the principal concerns and controlling factors for Arsenic contamination in Chile. *Sci. Rep.* **2023**, *13*, 11168. [CrossRef]
10. Osuna-Martínez, C.C.; Armienta, M.A.; Bergés-Tiznado, M.E.; Páez-Osuna, F. Arsenic in waters, soils, sediments, and biota from Mexico: An environmental review. *Sci. Total Environ.* **2021**, *752*, 142062. [CrossRef]
11. Ramos, W.; Ortega-Loayza, A.G.; Díaz, J.; de La Cruz-Vargas, J.A.; Tello, M.; Ronceros, G.; Loayza, M.; Gutierrez, E.L. Arsenicism by Chronic Exposure to Mine Tailings in Peru: An Analysis of 17 Cases with Lesions on Skin and/or Annexes. *Clin. Cosmet. Investig. Dermatol.* **2022**, *15*, 2407–2414. [CrossRef] [PubMed]
12. Custodio, M.; Cuadrado, W.; Peñaloza, R.; Montalvo, R.; Ochoa, S.; Quispe, J. Human Risk from Exposure to Heavy Metals and Arsenic in Water from Rivers with Mining Influence in the Central Andes of Peru. *Water* **2020**, *12*, 1946. [CrossRef]
13. Custodio, M.; Peñaloza, R. Evaluation of the Distribution of Heavy Metals and Arsenic in Inland Wetlands (Peru) Using Multivariate Statistical Methods. *Ecol. Eng. Environ. Technol.* **2021**, *22*, 104–111. [CrossRef]
14. Mamani Lopez, F.P.; Chambilla Quispe, V. Análisis Fisicoquímico Y Biológico Del Agua Para Consumo Humano En El Distrito De Huanuara. *Cienc. Desarro.* **2019**, *11*, 67–70. [CrossRef]
15. World Health Organization. *Arsenic in Drinking-Water: Background Document for Development of WHO Guidelines for Drinking-Water Quality*; World Health Organization: Geneva, Switzerland, 2019.
16. Abernathy, C.O.; Liu, Y.P.; Longfellow, D.; Aposhian, H.V.; Beck, B.; Fowler, B.; Goyer, R.; Menzer, R.; Rossman, T.; Thompson, C.; et al. Arsenic: Health effects, mechanisms of actions, and research issues. *Environ. Health Perspect.* **1999**, *107*, 593–597. [CrossRef]
17. Chen, Y.; Parvez, F.; Gamble, M.; Islam, T.; Ahmed, A.; Argos, M.; Graziano, J.H.; Ahsan, H. Arsenic exposure at low-to-moderate levels and skin lesions, arsenic metabolism, neurological functions, and biomarkers for respiratory and cardiovascular diseases: Review of recent findings from the Health Effects of Arsenic Longitudinal Study (HEALS) in Bangladesh. *Toxicol. Appl. Pharmacol.* **2009**, *239*, 184–192. [CrossRef]
18. McDermott, T.R.; Stolz, J.F.; Oremland, R.S. Arsenic and the gastrointestinal tract microbiome. *Environ. Microbiol. Rep.* **2020**, *12*, 136–159. [CrossRef]
19. Vahidnia, A.; van der Voet, G.B.; de Wolff, F.A. Arsenic neurotoxicity—A review. *Hum. Exp. Toxicol.* **2007**, *26*, 823–832. [CrossRef]
20. Mayer, J.E.; Goldman, R.H. Arsenic and skin cancer in the USA: The current evidence regarding arsenic-contaminated drinking water. *Int. J. Dermatol.* **2016**, *55*, e585–e591. [CrossRef]
21. Vineetha, V.P.; Raghu, K.G. An Overview on Arsenic Trioxide-Induced Cardiotoxicity. *Cardiovasc. Toxicol.* **2019**, *19*, 105–119. [CrossRef]
22. Kile, M.L.; Mazumdar, M. Arsenic and developmental toxicity and reproductive disorders. In *Handbook of Arsenic Toxicology*; Elsevier: Amsterdam, The Netherlands, 2023; pp. 593–605. ISBN 9780323898478.
23. Korte, N.E.; Fernando, Q. A review of arsenic (III) in groundwater. *Crit. Rev. Environ. Control* **1991**, *21*, 1–39. [CrossRef]
24. Bissen, M.; Frimmel, F.H. Arsenic—A Review. Part II: Oxidation of Arsenic and its Removal in Water Treatment. *Acta Hydrochim. Hydrobiol.* **2003**, *31*, 97–107. [CrossRef]
25. Kowalski, K.P. Advanced Arsenic Removal Technologies Review. In *Chemistry of Advanced Environmental Purification Processes of Water*; Elsevier: Amsterdam, The Netherlands, 2014; pp. 285–337. ISBN 9780444531780.
26. Ng, K.-S.; Ujang, Z.; Le-Clech, P. Arsenic Removal Technologies for Drinking Water Treatment. *Re/Views Environ. Sci. Bio/Technol.* **2004**, *3*, 43–53. [CrossRef]
27. Carneiro, M.A.; Pintor, A.M.A.; Boaventura, R.A.R.; Botelho, C.M.S. Current Trends of Arsenic Adsorption in Continuous Mode: Literature Review and Future Perspectives. *Sustainability* **2021**, *13*, 1186. [CrossRef]

28. Frankenberger, J.W.T. *Environmental Chemistry of Arsenic*; CRC Press: Boca Raton, FL, USA, 2001; ISBN 9780429180071.
29. Mohan, D.; Pittman, C.U. Arsenic removal from water/wastewater using adsorbents—A critical review. *J. Hazard. Mater.* **2007**, *142*, 1–53. [CrossRef] [PubMed]
30. Guan, X.; Du, J.; Meng, X.; Sun, Y.; Sun, B.; Hu, Q. Application of titanium dioxide in arsenic removal from water: A review. *J. Hazard. Mater.* **2012**, *215–216*, 1–16. [CrossRef] [PubMed]
31. Ashraf, S.; Siddiqa, A.; Shahida, S.; Qaisar, S. Titanium-based nanocomposite materials for arsenic removal from water: A review. *Heliyon* **2019**, *5*, e01577. [CrossRef]
32. Gamarra, F.; Medina, J.; Lanchipa, W.; Tamayo, R.; Sacari, E. Structural, Optical, and Arsenic Removal Properties of Sol-Gel Synthesized Fe-Doped TiO_2 Nanoparticles. *Nanomaterials* **2022**, *12*, 3402. [CrossRef]
33. Das, T.K.; Bezbaruah, A.N. Comparative study of arsenic removal by iron-based nanomaterials: Potential candidates for field applications. *Sci. Total Environ.* **2021**, *764*, 142914. [CrossRef]
34. Shan, H.; Mo, H.; Liu, Y.; Zeng, C.; Peng, S.; Zhan, H. As(III) removal by a recyclable granular adsorbent through dopping Fe-Mn binary oxides into graphene oxide chitosan. *Int. J. Biol. Macromol.* **2023**, *237*, 124184. [CrossRef]
35. Opiso, E.M.; Tabelin, C.B.; Ramos, L.M.; Gabiana, L.J.R.; Banda, M.H.T.; Delfinado, J.R.Y.; Orbecido, A.H.; Zoleta, J.B.; Park, I.; Arima, T.; et al. Development of a three-step approach to repurpose nickel-laterite mining waste into magnetite adsorbents for As(III) and As(V) removal: Synthesis, characterization and adsorption studies. *J. Environ. Chem. Eng.* **2023**, *11*, 108992. [CrossRef]
36. Tamayo, R.; Espinoza-González, R.; Gracia, F.; Rodrigues-Filho, U.P.; Flores, M.; Sacari, E. As(III) Removal from Aqueous Solution by Calcium Titanate Nanoparticles Prepared by the Sol Gel Method. *Nanomaterials* **2019**, *9*, 733. [CrossRef]
37. Shanmugaraj, K.; Vinoth, V.; Pugazhenthiran, N.; Valdés, H.; Salvo, C.; Sepúlveda, E.; Viswanathan Mangalaraja, R. Ferrihydrite−Graphene Oxide Foams as an Efficient Adsorbent for Arsenic(III) Removal from an Aqueous Solution. *Inorg. Chem. Commun.* **2023**, *153*, 110892. [CrossRef]
38. Singh, D.K.; Mohan, S.; Kumar, V.; Hasan, S.H. Kinetic, isotherm and thermodynamic studies of adsorption behaviour of CNT/CuO nanocomposite for the removal of As(III) and As(V) from water. *RSC Adv.* **2016**, *6*, 1218–1230. [CrossRef]
39. Siddiqui, S.I.; Naushad, M.; Chaudhry, S.A. Promising prospects of nanomaterials for arsenic water remediation: A comprehensive review. *Process Saf. Environ. Prot.* **2019**, *126*, 60–97. [CrossRef]
40. Massoudinejad, M.; Keramati, H.; Ghaderpoori, M. Investigation of photo-catalytic removal of arsenic from aqueous solutions using UV/H_2O_2 in the presence of ZnO nanoparticles. *Chem. Eng. Commun.* **2020**, *207*, 1605–1615. [CrossRef]
41. Reddy, K.J.; McDonald, K.J.; King, H. A novel arsenic removal process for water using cupric oxide nanoparticles. *J. Colloid Interface Sci.* **2013**, *397*, 96–102. [CrossRef] [PubMed]
42. Samad, A.; Furukawa, M.; Katsumata, H.; Suzuki, T.; Kaneco, S. Photocatalytic oxidation and simultaneous removal of arsenite with CuO/ZnO photocatalyst. *J. Photochem. Photobiol. A Chem.* **2016**, *325*, 97–103. [CrossRef]
43. Vaiano, V.; Iervolino, G.; Rizzo, L. Cu-doped ZnO as efficient photocatalyst for the oxidation of arsenite to arsenate under visible light. *Appl. Catal. B Environ.* **2018**, *238*, 471–479. [CrossRef]
44. Malwal, D.; Gopinath, P. Rapid and efficient removal of arsenic from water using electrospun CuO–ZnO composite nanofibers. *RSC Adv.* **2016**, *6*, 115021–115028. [CrossRef]
45. Savage, N.; Diallo, M.S. Nanomaterials and Water Purification: Opportunities and Challenges. *J. Nanoparticle Res* **2005**, *7*, 331–342. [CrossRef]
46. Khan, Q.U.; Begum, N.; Rehman, Z.U.; Khan, A.U.; Tahir, K.; Tag El Din, E.S.M.; Alothman, A.A.; Habila, M.A.; Liu, D.; Bocchetta, P.; et al. Development of Efficient and Recyclable ZnO-CuO/g-C_3N_4 Nanocomposite for Enhanced Adsorption of Arsenic from Wastewater. *Nanomaterials* **2022**, *12*, 3984. [CrossRef] [PubMed]
47. Penke, Y.K.; Kar, K.K. A review on multi-synergistic transition metal oxide systems towards arsenic treatment: Near molecular analysis of surface-complexation (synchrotron studies/modeling tools). *Adv. Colloid Interface Sci.* **2023**, *314*, 102859. [CrossRef] [PubMed]
48. Lata, S.; Samadder, S.R. Removal of arsenic from water using nano adsorbents and challenges: A review. *J. Environ. Manag.* **2016**, *166*, 387–406. [CrossRef] [PubMed]
49. Khan, Q.U.; Begum, N.; Khan, A.U.; Nazir, S.; Ali, A.; Tahir, K.; Aizaz Ud Din, M.; Afzal, A.M.; Liu, D.; Zhan, Y. Nanocomposite Based Adsorbent for Enhanced Arsenic Removal: Determination of Adsorption Mechanism, Kinetic Study, and Factors Affecting the Adsorption Capability of Nanomaterials. *SSRN J.* **2022**. [CrossRef]
50. Kumar, I.; Ranjan, P.; Quaff, A.R. Cost-effective synthesis and characterization of CuO NPs as a nanosize adsorbent for As (III) remediation in synthetic arsenic-contaminated water. *J. Environ. Health Sci. Eng.* **2020**, *18*, 1131–1140. [CrossRef]
51. Bekru, A.G.; Tufa, L.T.; Zelekew, O.A.; Goddati, M.; Lee, J.; Sabir, F.K. Green Synthesis of a CuO-ZnO Nanocomposite for Efficient Photodegradation of Methylene Blue and Reduction of 4-Nitrophenol. *ACS Omega* **2022**, *7*, 30908–30919. [CrossRef]
52. Mubeen, K.; Irshad, A.; Safeen, A.; Aziz, U.; Safeen, K.; Ghani, T.; Khan, K.; Ali, Z.; ul Haq, I.; Shah, A. Band structure tuning of ZnO/CuO composites for enhanced photocatalytic activity. *J. Saudi Chem. Soc.* **2023**, *27*, 101639. [CrossRef]
53. Renuka, L.; Anantharaju, K.S.; Vidya, Y.S.; Nagaswarupa, H.P.; Prashantha, S.C.; Nagabhushana, H. Synthesis of Sunlight Driven ZnO/CuO Nanocomposite: Characterization, Optical, Electrochemical and Photocatalytic Studies. *Mater. Today Proc.* **2017**, *4*, 11782–11790. [CrossRef]
54. Sakib, A.; Masum, S.; Hoinkis, J.; Islam, R.; Molla, M. Synthesis of CuO/ZnO Nanocomposites and Their Application in Photodegradation of Toxic Textile Dye. *J. Compos. Sci.* **2019**, *3*, 91. [CrossRef]

55. Gholami, P.; Dinpazhoh, L.; Khataee, A.; Orooji, Y. Sonocatalytic activity of biochar-supported ZnO nanorods in degradation of gemifloxacin: Synergy study, effect of parameters and phytotoxicity evaluation. *Ultrason. Sonochem.* **2019**, *55*, 44–56. [CrossRef]
56. Khataee, A.; Kalderis, D.; Gholami, P.; Fazli, A.; Moschogiannaki, M.; Binas, V.; Lykaki, M.; Konsolakis, M. Cu_2O-CuO@biochar composite: Synthesis, characterization and its efficient photocatalytic performance. *Appl. Surf. Sci.* **2019**, *498*, 143846. [CrossRef]
57. Gupta, K.; Joshi, P.; Gusain, R.; Khatri, O.P. Recent advances in adsorptive removal of heavy metal and metalloid ions by metal oxide-based nanomaterials. *Coord. Chem. Rev.* **2021**, *445*, 214100. [CrossRef]
58. Ayub, A.; Raza, Z.A.; Majeed, M.I.; Tariq, M.R.; Irfan, A. Development of sustainable magnetic chitosan biosorbent beads for kinetic remediation of arsenic contaminated water. *Int. J. Biol. Macromol.* **2020**, *163*, 603–617. [CrossRef] [PubMed]
59. Winiarski, J.; Tylus, W.; Winiarska, K.; Szczygieł, I.; Szczygieł, B. XPS and FT-IR Characterization of Selected Synthetic Corrosion Products of Zinc Expected in Neutral Environment Containing Chloride Ions. *J. Spectrosc.* **2018**, *2018*, 2079278. [CrossRef]
60. Da Silva-Neto, M.L.; de Oliveira, M.C.A.; Dominguez, C.T.; Lins, R.E.M.; Rakov, N.; de Araújo, C.B.; Menezes, L.d.S.; de Oliveira, H.P.; Gomes, A.S.L. UV random laser emission from flexible ZnO-Ag-enriched electrospun cellulose acetate fiber matrix. *Sci. Rep.* **2019**, *9*, 11765. [CrossRef]
61. Srivastava, R.; Anu Prathap, M.U.; Kore, R. Morphologically controlled synthesis of copper oxides and their catalytic applications in the synthesis of propargylamine and oxidative degradation of methylene blue. *Colloids Surf. A Physicochem. Eng. Asp.* **2011**, *392*, 271–282. [CrossRef]
62. Ramimoghadam, D.; Hussein, M.Z.B.; Taufiq-Yap, Y.H. Synthesis and characterization of ZnO nanostructures using palm olein as biotemplate. *Chem. Cent. J.* **2013**, *7*, 71. [CrossRef]
63. Phoohinkong, W.; Foophow, T.; Pecharapa, W. Synthesis and characterization of copper zinc oxide nanoparticles obtained via metathesis process. *Adv. Nat. Sci. Nanosci. Nanotechnol.* **2017**, *8*, 35003. [CrossRef]
64. Dhiman, V.; Kondal, N. ZnO Nanoadsorbents: A potent material for removal of heavy metal ions from wastewater. *Colloid Interface Sci. Commun.* **2021**, *41*, 100380. [CrossRef]
65. Hou, S.; Ding, W.; Liu, S.; Zheng, H.; Zhai, J.; Yang, L.; Zhong, Z. Fast oxidation and deep removal of As(III) by integrating metal–organic framework ZIF-67 and sulfite: Performance and mechanism. *Chem. Eng. J.* **2023**, *460*, 141785. [CrossRef]
66. Moussout, H.; Ahlafi, H.; Aazza, M.; Maghat, H. Critical of linear and nonlinear equations of pseudo-first order and pseudo-second order kinetic models. *Karbala Int. J. Mod. Sci.* **2018**, *4*, 244–254. [CrossRef]
67. Ho, Y.-S. Review of second-order models for adsorption systems. *J. Hazard. Mater.* **2006**, *136*, 681–689. [CrossRef] [PubMed]
68. Plazinski, W.; Rudzinski, W.; Plazinska, A. Theoretical models of sorption kinetics including a surface reaction mechanism: A review. *Adv. Colloid Interface Sci.* **2009**, *152*, 2–13. [CrossRef] [PubMed]
69. Ho, Y.; McKay, G. Pseudo-second order model for sorption processes. *Process Biochem.* **1999**, *34*, 451–465. [CrossRef]
70. Verma, L.; Singh, J. Arsenic adsorption from aqueous solution and groundwater using monometallic (Fe) and bimetallic (Fe/Mn) Tectona biochar synthesized from plant refuse: Mechanism, isotherm, and kinetic study. *Environ. Eng. Res.* **2023**, *28*, 220110. [CrossRef]
71. Banerjee, S.; Sharma, Y.C. Synthesis and application of Zn/Ce bimetallic oxides for the decontamination of arsenite (As-III) ions from aqueous solutions. *J. Environ. Manag.* **2019**, *233*, 151–164. [CrossRef] [PubMed]
72. Gao, X.; Root, R.A.; Farrell, J.; Ela, W.; Chorover, J. Effect of silicic acid on arsenate and arsenite retention mechanisms on 6-L ferrihydrite: A spectroscopic and batch adsorption approach. *Appl. Geochem.* **2013**, *38*, 110–120. [CrossRef]
73. Pratt, A. Environmental Applications of Magnetic Nanoparticles. In *Nanomagnetism: Fundamentals and Applications*; Elsevier: Amsterdam, The Netherlands, 2014; pp. 259–307. ISBN 9780080983530.
74. Brdar, M.; Takaci, A.; Sciban, M.; Rakic, D. Isotherms for the adsorption of Cu(II) onto lignin: Comparison of linear and non-linear methods. *Hem. Ind.* **2012**, *66*, 497–503. [CrossRef]
75. Manning, B.A.; Goldberg, S. Adsorption and Stability of Arsenic(III) at the Clay Mineral—Water Interface. *Environ. Sci. Technol.* **1997**, *31*, 2005–2011. [CrossRef]
76. Xie, M.; Luo, X.; Liu, C.; You, S.; Rad, S.; He, H.; Huang, Y.; Tu, Z. Enhancing mechanism of arsenic(III) adsorption by MnO2-loaded calcined MgFe layered double hydroxide. *RSC Adv.* **2022**, *12*, 25833–25843. [CrossRef] [PubMed]
77. Alam, M.A.; Shaikh, W.A.; Alam, M.O.; Bhattacharya, T.; Chakraborty, S.; Show, B.; Saha, I. Adsorption of As (III) and As (V) from aqueous solution by modified *Cassia fistula* (golden shower) biochar. *Appl. Water Sci.* **2018**, *8*, 198. [CrossRef]
78. Cai, G.; Tian, Y.; Li, L.; Zhang, W.; Huang, R.; Zhang, J.; Wang, Q.; Xu, H.; Zhang, Y. Sulfite activation of Fe-Mn bimetallic oxides for rapid oxidative removal of As(III) in water: Involvement of active Mn(III). *Chem. Eng. J.* **2024**, *479*, 147539. [CrossRef]
79. Yang, Y.; Zhang, R.; Chen, S.; Zhu, J.; Wu, P.; Huang, J.; Qi, S. Arsenic(iii) removal from aqueous solution using TiO_2-loaded biochar prepared by waste Chinese traditional medicine dregs. *RSC Adv.* **2022**, *12*, 7720–7734. [CrossRef] [PubMed]
80. Navarathna, C.M.; Karunanayake, A.G.; Gunatilake, S.R.; Pittman, C.U.; Perez, F.; Mohan, D.; Mlsna, T. Removal of Arsenic(III) from water using magnetite precipitated onto Douglas fir biochar. *J. Environ. Manag.* **2019**, *250*, 109429. [CrossRef] [PubMed]
81. Morales, H.M.; Torreblanca, G.; Mar, A.; Alcoutlabi, M.; Eubanks, T.M.; Plata, E.; Parsons, J.G. Investigation of the Thermodynamics for the Removal of As(III) and As(V) from Water Using Synthesized ZnO Nanoparticles and the Effects of pH, Temperature, and Time. *Appl. Sci.* **2023**, *13*, 10525. [CrossRef]
82. Singh, S.; Khasnabis, S.; Anil, A.G.; Kumar, V.; Kumar Naik, T.S.; Nath, B.; Garg, V.K.; Singh, J.; Ramamurthy, P.C. Multifunctional nanohybrid for simultaneous detection and removal of Arsenic(III) from aqueous solutions. *Chemosphere* **2022**, *289*, 133101. [CrossRef]

83. Khan, A.U.; Tahir, K.; Khan, Q.U.; Albalawi, K.; Jevtovic, V.; Almarhoon, Z.M.; El-Zahhar, A.A.; Al-Shehri, H.S.; Ullah, S.; Khan, M.U. Scaled-up development of recyclable Pd@ZnO/CuO nanostructure for efficient removal of arsenic from wastewater. *J. Mol. Struct.* **2022**, *1260*, 132828. [CrossRef]
84. Purohit, S.; Chini, M.K.; Chakraborty, T.; Yadav, K.L.; Satapathi, S. Rapid removal of arsenic from water using metal oxide doped recyclable cross-linked chitosan cryogel. *SN Appl. Sci.* **2020**, *2*, 768. [CrossRef]
85. Yuvaraja, G.; Prasad, C.; Vijaya, Y.; Subbaiah, M.V. Application of ZnO nanorods as an adsorbent material for the removal of As(III) from aqueous solution: Kinetics, isotherms and thermodynamic studies. *Int. J. Ind. Chem.* **2018**, *9*, 17–25. [CrossRef]
86. Qiu, Z.; He, X.; Qiu, F.; Yang, D.; Yao, G.; Zhang, T. Efficient oxidation and absorption of As(III) from aqueous solutions for environmental remediation via CuO@MNW membranes. *Sep. Purif. Technol.* **2020**, *250*, 117165. [CrossRef]
87. Othmani, A.; Maitlo, H.A.; Hamimed, S. Factors affecting adsorption capabilities of nanoscale materials. In *Adsorption through Advanced Nanoscale Materials*; Elsevier: Amsterdam, The Netherlands, 2023; pp. 47–64. ISBN 9780443184567.
88. Singh, P.; Sarswat, A.; Pittman, C.U.; Mlsna, T.; Mohan, D. Sustainable Low-Concentration Arsenite As(III) Removal in Single and Multicomponent Systems Using Hybrid Iron Oxide-Biochar Nanocomposite Adsorbents-A Mechanistic Study. *ACS Omega* **2020**, *5*, 2575–2593. [CrossRef] [PubMed]
89. Reddad, Z.; Gerente, C.; Andres, Y.; Le Cloirec, P. Adsorption of several metal ions onto a low-cost biosorbent: Kinetic and equilibrium studies. *Environ. Sci. Technol.* **2002**, *36*, 2067–2073. [CrossRef] [PubMed]
90. Kanematsu, M.; Young, T.M.; Fukushi, K.; Green, P.G.; Darby, J.L. Arsenic(III, V) adsorption on a goethite-based adsorbent in the presence of major co-existing ions: Modeling competitive adsorption consistent with spectroscopic and molecular evidence. *Geochim. Cosmochim. Acta* **2013**, *106*, 404–428. [CrossRef]
91. Tuutijärvi, T.; Repo, E.; Vahala, R.; Sillanpää, M.; Chen, G. Effect of Competing Anions on Arsenate Adsorption onto Maghemite Nanoparticles. *Chin. J. Chem. Eng.* **2012**, *20*, 505–514. [CrossRef]
92. Jain, A.; Loeppert, R.H. Effect of Competing Anions on the Adsorption of Arsenate and Arsenite by Ferrihydrite. *J. Environ. Qual.* **2000**, *29*, 1422–1430. [CrossRef]

water

MDPI

Article

Control of Bromate Formation in Desalinated Seawater Production and Transmission with Ammoniation

Ali A. Alhamzah [1,2], Abdulrahman S. Alofi [3], Abdulrahman A. Abid [3] and Christopher M. Fellows [1,2,*]

[1] Water Technologies Innovation Institute & Research Advancement (WTIIRA), Saline Water Conversion Corporation, Jubail 31951, Saudi Arabia; aal-hamzah2@swcc.gov.sa
[2] School of Science and Technology, The University of New England, Armidale, NSW 2351, Australia
[3] Saline Water Conversion Corporation, Riyadh 11547, Saudi Arabia; aalofi@swcc.gov.sa (A.S.A.); aabid@swcc.gov.sa (A.A.A.)
* Correspondence: cfellows@une.edu.au or cmichael@swcc.gov.sa; Tel.: +966-59-417-1150

Abstract: Bromate is a potentially carcinogenic disinfection by-product of potential concern in desalinated waters, where bromide derived from seawater can be converted to bromate by the oxidising species used for disinfection. Historically, it has been difficult to maintain complete adherence to national standards of no more than 10 ppb for bromate at all locations served with desalinated seawater by the Saline Water Conversion Corporation (SWCC) in the Kingdom of Saudi Arabia. In this full-scale study, the addition of 100–200 ppb of ammonia to the produced water of a Multi-Stage Flash Desalination plant effectively controlled the formation of bromate in the transmission system supplying inland centres in the Makkah Province of the Kingdom of Saudi Arabia (Arafa, Taif) on a time scale sufficient for the distribution of water to the consumer, even when the bromide content of the produced water was artificially enhanced (up to 132 ppb) via the addition of seawater.

Keywords: ammonia; bromate; desalination; multi-stage flash; transmission

1. Introduction

A significant public health concern in any system where potable water is stored or transported is the possibility of the formation of disinfection by-products. Disinfection is a necessity to avoid the bacterial contamination of drinking water, but the oxidising species used can generate a range of potentially toxic and carcinogenic species from trace components of the produced water [1–4]. When water contains bromide ions, these disinfection by-products can include the bromate ion as well as brominated organic species such as bromoacetic acid, bromoform, and bromodichloromethane [5,6]. Historically, bromate control has been an issue primarily in surface and ground water treatment with ozonation, and there is a considerable body of research literature addressing control measures for this problem [7–15].

Recent interest in bromate control in the Kingdom of Saudi Arabia in waters treated by chlorination rather than ozonation arises from a unique combination of two factors: (1) increasing use of seawater desalination by reverse osmosis (SWRO), rather than thermal methods, which leads to product water with a higher concentration of bromide; (2) transmission of the majority of produced water over lengthy pipelines (>100 km) at relatively high temperatures, as the formation of disinfection by-products increases with both temperature and time. While SWRO is a well-established technology in many parts of the world, outside of the Kingdom of Saudi Arabia it is used almost entirely to serve coastal centres, where it is consumed in close proximity to its point of production. These two factors have led to challenges in consistently meeting the 10 ppb maximum limit set by regulatory authorities in the Kingdom of Saudi Arabia [16,17].

This work reports on efforts by the Saline Water Conversion Corporation (SWCC) to minimize the formation of bromate and related disinfection by-products in the water transmitted to consumers via the addition of ammonia to produced desalinated water. We will first review and discuss the mechanism of bromate formation in water treated by chlorination and the probable role of ammonia in affecting this mechanism will in the light of SWCC's experience in monitoring water quality. The details of the application of ammonia at a SWCC desalination plant at Shoiabah on the Red Sea and the results of this application will then be presented.

2. Mechanisms and Kinetics

The mechanism of the formation of bromate under chlorination is well understood, with the first step being the formation of hypochlorite.

$$Cl_2 + H_2O \rightarrow HClO + H^+ + Cl^- \quad (1)$$

$$Cl_2 + H_2O \rightarrow ClO^- + 2H^+ + Cl^- \quad (2)$$

At the temperatures of interest in the storage and transmission of product water in the Kingdom of Saudi Arabia (15–45 °C), reaction (2) is dominant, and most of the dissolved oxidant is hypochlorite, rather than hypochlorous acid [18,19].

The second step is the oxidation of bromide to hypobromite under basic (reaction (3)) or acidic (reaction (4)) conditions.

$$ClO^- + Br^- \rightarrow BrO^- + Cl^- \quad (3)$$

$$HClO + Br^- \rightarrow HBrO + Cl^- \quad (4)$$

Both reactions are thermodynamically favourable over the temperature range of interest. However, the reaction rate for the reaction under acidic conditions is about 10^6 times higher, and the reaction is acid-catalyzed [20]. It is intuitive that this should be the more favourable reaction because it does not involve bringing two anions together. Thus, the reaction will be dominated by reaction (4) even at relatively high pH and the reaction under basic conditions can be ignored over the pH range of interest.

The third step is the oxidation of hypobromite to bromate. There is a significant body of literature on mechanisms for this reaction under ozonation conditions, but relatively little work has been conducted on the mechanism in the presence of an oxidant. Stoichiometrically, this is reported to occur by the disproportionation of hypobromous acid to give bromate and bromide [21,22]:

$$3HBrO \rightarrow HBrO_3 + 2HBr \quad (5)$$

Trimolecular reactions are statistically implausible, so this process is likely to proceed in two steps, for example, as proposed by Margerum and Huff Hartz [23]:

$$2HOBr \rightarrow BrO_2^- + Br^- + 2H^+ \ (pK_a\ HBrO_2 = 6.25) \quad (6)$$

Followed by:

$$HOBr + BrO_2^- \rightarrow BrO_3^- + Br^- + H^+ \ (pK_a\ HBrO_3 = -2) \quad (7)$$

A similar second step for the formation of bromate from hypobromite has been previously reported with ozone as an oxidant [24]. Margerum and Huff Hartz found that reaction (5) was second order in HOBr in the presence of HOCl, and thus proposed [23]:

$$2HOBr \rightarrow BrO_2^- + Br^- + 2H^+ \ k = 0.015\ M^{-1}s^{-1} \quad (8)$$

$$HOCl + BrO_2^- \rightarrow BrO_3^- + Cl^- + H^+ \; k = \text{fast} \tag{9}$$

with the first step clearly being rate limiting. The apparent rate coefficient k increased with total chlorine, but this only became evident above 5 mM (>175 ppm) Cl; Margerum and Huff Hartz reported that the rate constant for the second-order decomposition of HOBr has a maximum at pH ~7.2 and fit this with a complex mechanistic model, but this maximum is dependent on only one data point at pH 7.6 [23]. In contrast to this, we have consistently observed more bromate formation at higher pH over the range of pH values seen in product water (7.5–9.0) in routine assessments of water quality in the SWCC production and transmission systems. Therefore, we postulate a significant contribution of the cross-reaction between hypobromite and hypobromous acid:

$$HOBr + BrO^- \rightarrow BrO_2^- + Br^- + H^+ \tag{10}$$

This reaction should be favourable since it requires less charge separation overall, and if it goes at a slower rate than the 2HOBr reaction, it could contribute negligibly at pH ~7 but give elevated rates any higher pHs, as observed in the SWCC transmission system, for example, if the HOBr + BrO^- reaction proceeds at one-third the rate of the 2HOBr reaction (Figure 1).

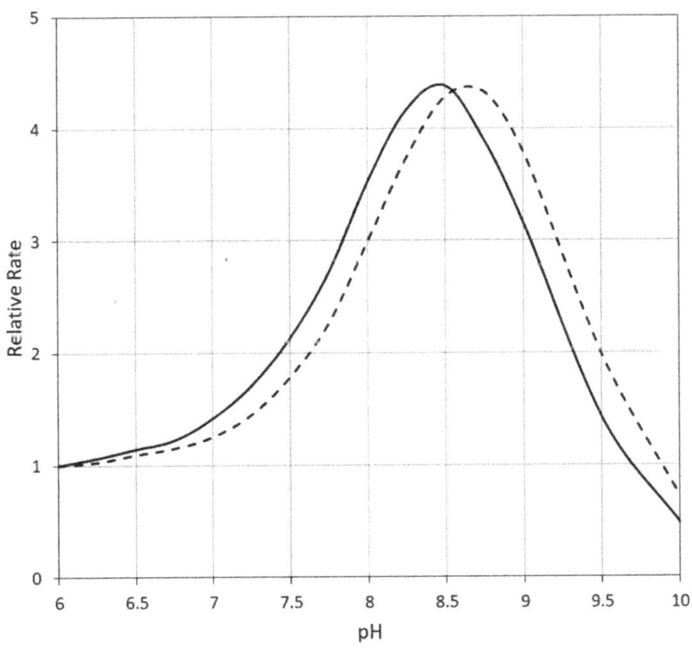

Figure 1. Modelled relative rate of the generation of hypobromate with pH at T = 20 °C (dashed line) and 40 °C (solid line), assuming the rate limiting step is the bimolecular reaction of HOBr with HOBr (relative k = 3) or HOBr with OBr^- (relative k = 1).

The overall scheme of bromate formation with estimated kinetic and thermodynamic parameters of importance is summarized below (Scheme 1).

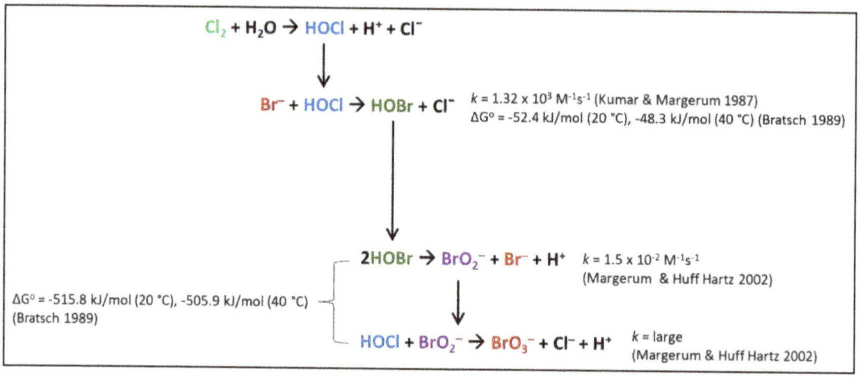

Scheme 1. Pathways of formation of bromate from bromide with chlorine oxidant [20,23,25].

Ammonia can initially impact the system as outlined above (Scheme 1) in two primary ways [24]:

(i) Reducing the amount of HOCl available for the oxidation of bromide [26]:

$$NH_3 + HOCl \rightarrow NH_2Cl + H_2O\ (k = 4.2 \times 10^6\ M^{-1}s^{-1}) \quad (11)$$

(ii) Reacting with HOBr so it is unavailable for oxidation to bromate [27]:

$$NH_3 + HOBr \rightarrow NH_2Br + H_2O\ (k = 7.5 \times 10^7\ M^{-1}s^{-1}) \quad (12)$$

Given the relative concentrations expected of HOCl and HOBr, reaction (11) is likely to be more important than reaction (12) despite the order of magnitude difference in rates. Because of the impact of ammonia on reducing the amount of HOCl (while not losing disinfectant capacity, as monochloroamine is also a disinfectant), Ling et al. recommended the addition of ammonia before chlorination in ozonation reactions [24].

These main reactions removing HOBr and hence reducing bromate formation are shown below (Scheme 2).

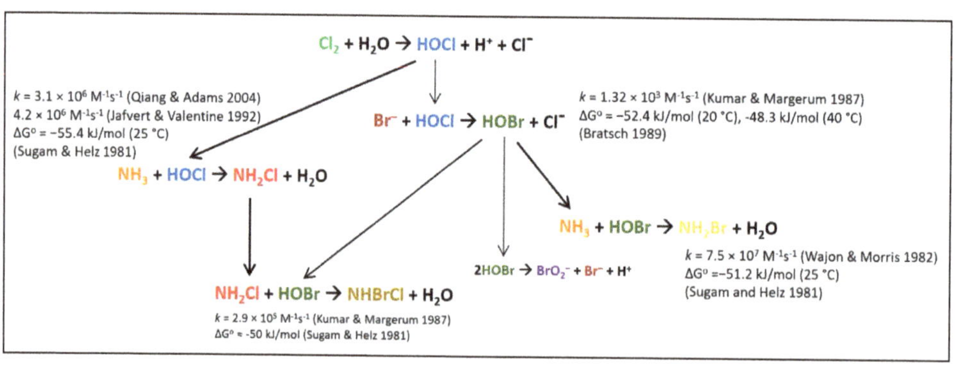

Scheme 2. Pathways for inhibition of bromate from bromide with chlorine oxidant [20,25,27–29].

The monochloramine and monobromamine can react further [30]:

$$NH_2Cl + HOBr \rightarrow NHBrCl + H_2O\ (k = 2.9 \times 10^5\ M^{-1}s^{-1}). \quad (13)$$

Meanwhile, NH_2Br may disproportionate reversibly in a base-catalyzed reaction [31]:

$$2NH_2Br \leftrightharpoons NHBr_2 + NH_3 \text{ (}K\text{ of order 0.5-5 depending on catalyst)} \quad (14)$$

Following that, it will then react with NH_2Cl [32]:

$$NHBr_2 + NH_2Cl \rightarrow NH_2Br + NHBrCl \quad (15)$$

These haloamines may also react with any remaining bromide present in the system, giving bromamine [33],

$$NH_2Cl + Br^- \rightarrow NH_2Br + Cl^- \ (k = 0.014\ M^{-1}s^{-1}) \quad (16)$$

bromochloramine [33,34],

$$2NH_2Cl + Br^- \rightarrow NHBrCl + NH_3 + Cl^- \ (k = 2.9\text{–}3.4 \times 10^6\ M^{-2}s^{-1}) \quad (17)$$

or dibromamine [24,35]:

$$NHBrCl + Br^- \rightarrow NHBr_2 + Cl^- \ (k = 565\ M^{-1}\ s^{-1}) \quad (18)$$

The net result of these reactions should be the sequestration of the initially present bromide as haloamines, predominantly as NHBrCl [33,36]. The net of reactions of the haloamines is summarized below (Scheme 3).

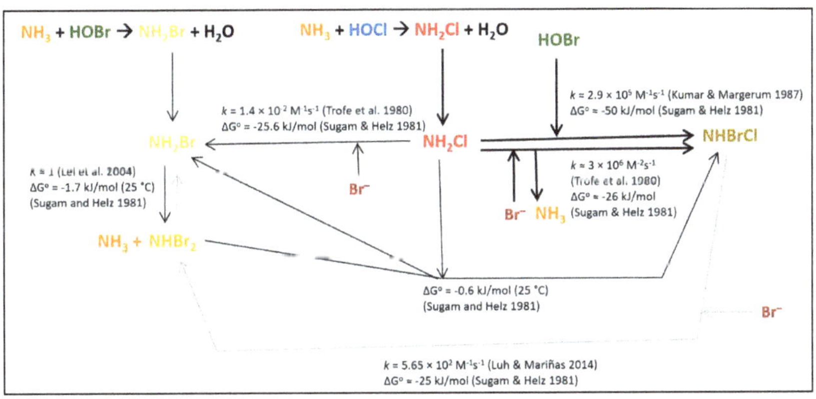

Scheme 3. Interaction of haloamine reactions in ammoniated and chlorinated waters containing bromide [20,29,31,33,35].

NHBrCl has been reported to be less reactive than the mono- and dibromamines [37], giving an expected impact on the yield of halogenated organic contaminants, as well as bromate (pace Valentine [38]). It is not wise to trust an experimental result until it has been validated by theory [39], and the experimental observation is supported by the free energies calculated for bromamines and chloramines in aqueous solution by Trogolo and Aret [40]. These free energy values suggest that for a generic bromination reaction,

$$A + NH_xBr_y(Cl) \rightarrow ABr + NH_{x-1}Br_{y-1}(Cl) \quad (19)$$

the free energy of the reaction will be higher for NHBrCl than for NH_2Br (2.5 kJ/mol less) and $NHBr_2$ (5.5 kJ/mol less). Thus, the ammoniation before chlorination process can be expected to strongly reduce the incidence of other brominated disinfection products. The addition of ammonia before chlorination will cause the formation of increased levels of chloramine. Ammoniation after chlorination is more problematic, because if the rate coefficients quoted above are correct, this will lead to a significant population of NH_2Br and $NHBr_2$, which are good brominating agents [41,42]. However, minimizing the time between chlorination and ammoniation to approximately 30 s has also been shown to give primarily NHBrCl [43]. Sun et al. reported in a similar system to desalination (secondary effluent chlorination) that trihalomethane (THM) production was reduced but haloacetic acid concentrations were increased at a bromide concentration of 140 ppb and not reduced at 50 ppb; it is unclear from the description of this study whether ammonia was added before or after chlorination [44]. This is consistent with other reports of reduced THM yields under conditions where bromamines replace hypobromous acid [29] and that bromamines are more reactive in forming haloacetic acids [41].

As long as there is residual chlorine in the system, it is expected that this chlorine will continue to generate dibromochloramine. Only when there is no remaining oxidant is NBr_2Cl expected to decompose in aqueous solution to H^+, N_2, Cl^-, Br^-, and BrO^- [35]. Luh and Mariñas (2014) studied the decomposition of NH_2Cl with a large excess (by 5 to 50 times) of Br^- over NH_2Cl and clearly observed the replacement of NH_2Cl by NHBrCl, which decomposed at a slower rate [35]. At the lower relative concentrations of bromide in desalination product water (~100 ppb), this suggests that the bromide-consuming reactions discussed here will not deplete the NH_2Cl available. Hu et al. calculated for a system also containing relatively high concentrations of bromide, but with an excess of chloramine over bromide (0.05 mM NH2Cl, 1.6–3.2 ppm 0.02/0.04 mM Br^-), that almost all bromine would be present as NHBrCl, and found it to be rapidly degraded by CuO [45].

The body of data obtained up until now and outlined here on the mechanism and kinetics of the interactions between ammonia, chlorine, and bromide therefore suggests that chlorination followed by ammoniation of the water produced by seawater desalination is a potential strategy for controlling the formation of bromate and other disinfection by-products of concern.

3. Materials and Methods

Water samples were collected from the Shoaibah Phase 2 desalination plant of the SWCC and from a number of storage tanks adjacent to the plant, pumping stations on transmission lines originating at the Shoaibah desalination complex, and storage tanks in communities near the termini of these transmission lines (Arafa and Taif), in 2022 and 2023. The production and transmission system is complex, and a simplified diagram showing the relationship between the different sampling points is given below (Figure 2).

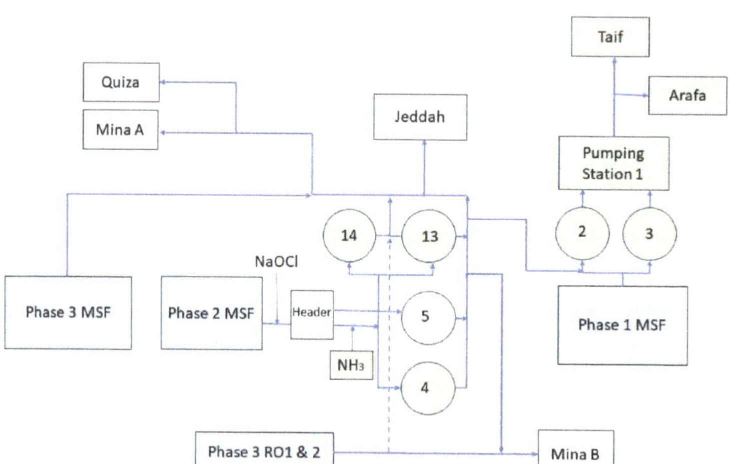

Figure 2. Sampling points in the Shoaibah desalination plant campus and associated transmission system.

Analysis of the parameters of interest was carried out within three weeks of collection in the laboratories of the Desalination Technologies Research Institute of the SWCC in Al Jubail, Kingdom of Saudi Arabia, using standard methods as follows:

pH: Potentiometry using a standard hydrogen electrode [46];

Ammonium, bromide, bromate, chloride, and nitrate: Ion chromatography with chemical suppression of eluent conductivity [47].

Trihalomethanes and other organic contaminants: EPA 524 (Gas chromatography-Mass spectrometry, GC-MS). One sample (5 May 2022) was analyzed in the WTIIRA laboratories; one sample (25 May 2023) by ALS Arabia, Dammam, Saudi Arabia; and one sample (6 June 2023) by SGS Inspection Services, Jubail, Saudi Arabia.

Ammonia was added to an existing post-treatment system utilising chlorination via sodium hypochlorite. The ammonia addition point was located about 50 m downstream of the chlorination point.

Two rounds of trials were made, in 2022 and 2023.

The 2022 trials were carried out in several stages:

(0) Addition of sufficient seawater to bring the TDS up to 82–94 ppm, which should have given bromide concentrations in the range 49–72 ppb if there was no selective rejection or permeation of bromide (14–16 April 2022);

(1) Addition of 120 ppb NH_3 along with sufficient seawater to bring the TDS up to 81–89 ppm, which should have given bromide concentrations in the range 47–64 ppb (17–19 April 2022);

(2) Addition of 120 ppb NH_3 along with sufficient seawater to bring the TDS up to 106–115 ppm, (96–113 ppb Br^-) (20–25 April 2022);

(3) Addition of 200 ppb NH_3, along with sufficient seawater to bring the TDS up to 110–132 ppm (104–132 ppb Br^-) (26 April–3 May 2022);

(4) Addition of 200 ppb NH_3 and sufficient sodium hydroxide to adjust the pH to 8.7, along with sufficient seawater to give a TDS of 114 ppm (112 ppb bromide) (4–5 May 2022).

For a number of locations, control data were also collected on 12–13 April 2022.

A second series of tests was carried out in 2023, with the application of chlorination/ammoniation in both Phase 1 and Phase 2 of the Shoaibah Desalination plant:

(1) Addition of 100 ppb NH_3 only (23 April–5 May 2023);

(2) Addition of 100 ppb NH_3 along with sufficient seawater to bring the TDS up to 119–242 ppm, which should have given bromide concentrations in the range 116–325 ppb. (4–14 May 2023);

(3) Addition of 100 ppb NH_3, along with sufficient seawater to bring the TDS up to 130–230 ppm, (134–305 ppb Br^-) (15 May–17 June 2023);

(4) Addition of 100 ppb NH_3 along with sufficient seawater to bring the TDS up to 82–125 ppm (52–126 ppb Br^-) (19 June–12 July 2023);

(5) Addition of 90 ppb NH_3 along with sufficient seawater to bring the TDS up to 82–129 ppm (52–133 ppb Br^-) (13 July–7 Aug 2023).

As each phase in the 2022 and 2023 test series corresponds to multiple measurements of a number of days, the average value and an error value calculated from the standard deviation in the measured values are reported for each measured parameter for each phase.

4. Results and Discussion

The key parameters of interest in both the 2022 and 2023 trials were the concentrations of bromide, bromate, ammonium, and nitrate in the product waters. Bromide and bromate were taken as the critical input and output of the complex system outlined in the introduction, and ammonium and nitrate were taken as potentially problematic by-products of the ammonia added to the system.

4.1. 2022 Trials

The measured bromide concentrations before chlorination and ammoniation were significantly below the predicted values, with 87 ± 4 ppb measured during stages 3 and 4. These were also below the value that could be predicted from measurements of the chloride concentration in the water (99 ± 3 ppb) using the expected ratio of Cl:Br in seawater. As the absolute number of bromine-containing species is less important than their relative proportions, the correlation between chloride and bromide concentration was used to estimate the total bromine concentration in product water samples after chlorination and compared to the amount of bromide.

After chlorination and ammoniation, the treated water was passed on to tanks within the Shoaibah desalination plant complex, which also receive water from other sources, some of which produce low-TDS water with no significant bromate problems, while others are more problematic; these in turn proceed to pumping stations which draw upon multiple tanks.

While the net of reactions described in Scheme 3 suggests that very little free bromide will be observed, being sequestered as bromamines if it is not present as hypobromite ion/hypobromous acid, ion chromatography found significant amounts of bromide ions in all waters investigated. In waters treated with ammonia, concentrations of bromide were consistently higher. Unless there are serious issues with the mechanisms postulated in the literature, this suggests that the bromamines are not stable under the conditions of ion chromatography. High levels of bromide in IC have also been observed by Pearce et al. in studies where high concentrations of chloramine were added to product water [48].

From Table 1, it can be seen that the apparent ratio of non-bromide bromine-containing species (presumably primarily BrO^-) as a fraction of the total bromine in the stream decreases on addition of 120 ppm ammonia (cf. the values obtained in Tank 5 with Tanks 4, 13, and 14), decreasing further as ammonia concentrations are increased to 200 ppm, and decreasing further again when pH is increased.

Table 1. Estimated % of non-bromide bromine at locations within the Shoaibah desalination plant site and associated transmission system, 2022. Values **bold and underlined** are locations where water quality should be affected by ammoniation.

Stage	Control	0	1	2	3	4
Line B after chlorination, before ammoniation		72 ± 4				
Header 1					58 ± 22	8 ± 4
Header 2					**18 ± 22**	**6 ± 6**
Tank 4		70 ± 2	**53 ± 10**	**21 ± 23**	**8 ± 4**	**0**
Tank 5		53 ± 12	61 ± 9	54 ± 20	43 ± 21	63 ± 3
Tank 13	73 ± 5	74 ± 2	**37 ± 28**	**48 ± 22**	**5 ± 2**	**2 ± 4**
Tank 14		67 ± 6	**44 ± 29**	**46 ± 26**	**5 ± 5**	**0**
PS (Pumping Station) 1A	0	4	7 ± 4	6 ± 1	3	0
PS1B	61 ± 4	43	**68 ± 7**	**55 ± 12**	**8 ± 10**	**0**
Jeddah PS	63 ± 1	44	61 ± 17	59 ± 17	56 ± 13	49
Quiza PS	47 ± 14	58	65 ± 13	70 ± 10	66 ± 11	23
Mina A PS	47 ± 27	53	49 ± 36	66 ± 12	35 ± 8	60
Mina B PS	46 ± 26	72	54 ± 12	61 ± 15	55 ± 17	78

From other indicators (F, SO_4, Ca) the water sampled in PS-1A was not derived from the water treated by chlorination–ammoniation during this time, but from the Phase 1 MSF plant (Figure 2). No residual NH_4 was measured in this water at any time during the study.

Note the clear fall in non-bromide bromine in Tanks 4, 13, and 14 and pumping station 1B over the time of the test: a small reduction at 100 ppb NH_3 treatment, followed by a larger fall at 200 ppb NH_3 and a larger fall again when pH was adjusted upwards.

Bromate reduction is not clearly significant with 120 ppb NH_3 addition (stages 1 and 2) but is marked at 200 ppb bromate addition (stages 3 and 4) (Table 2).

Table 2. Measured bromate ion concentration (ppb) at locations within the Shoiabah desalination plant site and associated transmission system, 2022. Values **bold and underlined** are locations where water quality should be affected by ammoniation.

Stage	Control	0	1	2	3	4
Line B after chlorination, before ammoniation		0				
Header 1					0	0
Header 2					**0**	**0**
Tank 4		7 ± 1	**5 ± 2**	**4**	**1 ± 1**	**0**
Tank 5		12 ± 4	7 ± 2	9 ± 2	14 ± 2	16 ± 1
Tank 13	2 ± 1	2 ± 1	**2 ± 1**	**2 ± 1**	**0**	**0**
Tank 14		2 ± 1	**2 ± 1**	**1 ± 1**	**0**	**0**
PS (Pumping Station) 1A	0	2	2	1 ± 1	0	0
PS1B	2	2	**2**	**1 ± 1**	**0**	**0**
Jeddah PS	3 ± 1	3	2	2	1	3 ± 1
Quiza PS	6 ± 1	4	6 ± 1	5 ± 2	5	4
Mina A PS	6 ± 1	4	5 ± 1	4 ± 1	6	6 ± 1
Mina B PS	5 ± 1	4	4 ± 1	2 ± 1	4 ± 1	5 ± 1

The day-by-day bromate data for the tanks containing untreated water (Tank 5) and primarily treated water (Tanks 4, 13, and 14) are displayed in Figure 3.

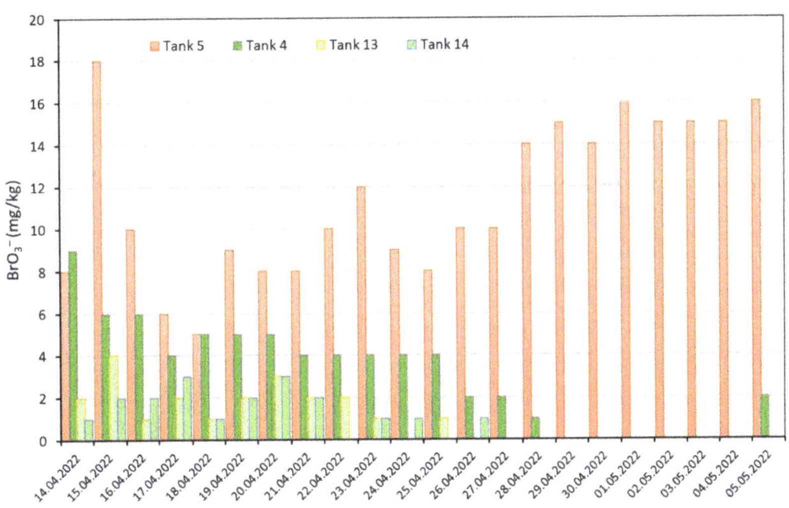

Figure 3. Bromate concentration in Shoiabah storage tanks during the 2022 trials.

Ammonium concentration should be a good indicator for the presence of treated water. It can be seen that there is essentially no ammonium at any of the pumping stations and disproportionately greater quantities of residual ammonium with 200 ppb treatment (Table 3). See, for example, Tank 4 at stage 2, where the effect of the treatment appears to be significant in terms of the bromide/hypobromite ratio and the amount of bromate observed. It is also clear that the analysis is not quantitative, as concentrations of ammonium up to twice the predicted concentration of the ammonia added are observed in some instances. There are also occasional outliers with no clear explanation at locations where ammonium should not be present, and an absence of ammonium under conditions (Tank 4, stages 1–2) where it would be expected; there are thus concerns about the methodology of ammonium determination.

Table 3. Measured ammonium ion concentration (ppb) at locations within the Shoiabah desalination plant site and associated transmission system, 2022. Values **bold and underlined** are locations where water quality should be affected by ammoniation.

Stage	Control	0	1	2	3	4
Line B after chlorination, before ammoniation		0				
Header 1					14 ± 23	0
Header 2					**282 ± 118**	**403 ± 107**
Tank 4		0	**0**	**0**	**162 ± 73**	**306 ± 10**
Tank 5		0	0	0	12 ± 31	0
Tank 13	0	0	**36 ± 29**	**6 ± 14**	**123 ± 58**	**180 ± 31**
Tank 14		0	**40 ± 32**	**24 ± 26**	**140 ± 60**	**200 ± 28**
PS (Pumping Station) 1A	0	0	0	0	0	62
PS1B	0	0	**0**	**16 ± 23**	**0**	**0**
Jeddah PS	20 ± 21	0	0	0	0	0
Quiza PS	0	0	0	0	0	0
Mina A PS	0	0	20 ± 30	0	0	0
Mina B PS	0	0	0	0	0	0

One concern about addition of ammonia is its potential oxidation to nitrate, with implications for the quality of water delivered to the consumer. The ppb of nitrate present in the system was also assessed (Table 4). There is no clear relationship between the amount of nitrate observed and the amount of ammonia added to the system. Although nitrate levels in Tank 5, not receiving the treated water, were assessed as lower than Tanks 4, 13, and 14 in phases 2 and 3, comparable values were obtained for all four tanks in phase 4.

Table 4. Measured nitrate ion concentration (ppb) at locations within the Shoiabah desalination plant site and associated transmission system, 2022. Values **bold and underlined** are locations where water quality should be affected by ammoniation.

Stage	Control	0	1	2	3	4
Line B after chlorination, before ammoniation		29 ± 4				
Header 1					32 ± 5	28
Header 2					**32 ± 4**	**25**
Tank 4		27 ± 1	**34 ± 4**	**40 ± 7**	**44 ± 5**	**39 ± 1**
Tank 5		16 ± 2	31 ± 14	24 ± 6	26 ± 6	31
Tank 13	35 ± 1	20 ± 4	**32 ± 10**	**39 ± 5**	**38 ± 3**	**28 ± 2**
Tank 14		25 ± 3	**35 ± 6**	**41 ± 11**	**39 ± 4**	**28 ± 1**
PS (Pumping Station) 1A	53 ± 29	14	38 ± 13	35 ± 3	42 ± 4	32
PS1B	29 ± 3	53	**27 ± 4**	**32 ± 5**	**37 ± 4**	**53**
Jeddah PS	31 ± 1	33	25 ± 1	34 ± 6	36 ± 4	30
Quiza PS	33	29	28 ± 1	29 ± 2	34 ± 1	33
Mina A PS	45 ± 14	30	34 ± 8	30 ± 1	35 ± 2	26
Mina B PS	9 ± 5	16	21 ± 5	15 ± 5	15 ± 3	19

The data on sites taking water not treated by this protocol illustrate the range in values found within the systems due to variations in the quality of the water produced, the mixing of these waters in different proportions, and the different environmental factors affecting these waters as they move through the transmission system.

Under the phase 4 trial conditions, samples of water from different parts of the Shoiabah plant were analyzed for trihalomethanes (Table 5). Trihalomethanes were only detected in the waters that were chlorinated and not ammoniated. Note that these species were most likely generated during the time when the samples were transported to the laboratory for analysis.

Table 5. Measured trihalomethanes concentration (ppb) at locations within the Shoiabah desalination plant site and associated transmission system on 5 May 2022.

Stage	$CHCl_3$	$CHCl_2Br$	$CHClBr_2$	$CHBr_3$
Line B before chlorination	<1	<1	<1	<1
Line B after chlorination, before ammoniation	3	9	18	13
Header 1	<1	<1	<1	<1
Tank 13	<1	<1	<1	<1
Tank 14	<1	<1	<1	<1

4.2. 2023 Trials

As in 2022, a significant reduction in non-bromide bromine was seen in 2023 in the tanks being fed by the Phase 2 Shoiabah desalination plant where the combined chlorination/ammoniation post-treatment was employed, as well as at in pumping station 1 (Table 6) Reductions in bromate at these tanks were also observed (Table 7). Data are given to compare water further down the transmission line not deriving from the treated water (at the PS Jeddah, Quiza, Mina A and Mina B) and also for water deriving from the treated water. The impact of the treatment on the proportion of non-bromide bromine appears to be clear at tanks (Arafa and Taif) located hundreds of kilometres from the water production site. While there is also a dramatic reduction in non-bromide bromine at Mina B pumping station between the control period, which did not receive the ammoniated water, taking the values in aggregate it is clear that this arises from an outlier in the upwards direction during the control period.

Table 6. Estimated % of non-bromide bromine at locations within the Shoiabah desalination plant site and associated transmission system, 2023. Values **bold and underlined** are locations where water quality should be affected by ammoniation.

Stage	Control	1	2	3	4
Phase 1–Tank 2	18 ± 31	**15 ± 21**	**11 ± 13**		
Phase 1–Tank 3	43 ± 11	**0 ± 11**	**0 ± 7**	**8 ± 8**	
Tank 4	36 ± 10	**3 ± 13**	**26 ± 19**	**35 ± 18**	
Tank 13	50 ± 13	**11 ± 9**	**7 ± 27**	**23 ± 25**	
Tank 14	42 ± 14	**3 ± 15**	**13 ± 19**	**30 ± 13**	
PS 1	49	**18 ± 14**	**24 ± 13**	**27 ± 16**	**8 ± 35**
Jeddah PS	25	39 ± 9	33 ± 23	34 ± 26	
Quiza PS	42	31 ± 9	28 ± 8	40 ± 23	
Mina A PS	50	54 ± 11	55 ± 12	61 ± 9	
Mina B PS	96	33 ± 6	28 ± 7	56 ± 5	
Arafa Tank Outlet	52 ± 17	**11 ± 7**	**12 ± 10**	**30 ± 26**	**36 ± 22**
Taif Tank Outlet	38 ± 9	**19 ± 3**	**20 ± 15**	**20 ± 15**	

Table 7. Measured bromate ion concentration (ppb) at locations within the Shoiabah desalination plant site and associated transmission system, 2023. Values **bold and underlined** are locations where water quality should be affected by ammoniation.

Stage	Control	1	2	3	4	5
Header–Phase 2					0	**0.17 ± 0.37**
Phase 1–Tank 2	1 ± 1	**1 ± 3**	0			
Phase 1–Tank 3	1 ± 1	0	0	0	0	0
Tank 4	3 ± 2	0	0	0		
Tank 13	2 ± 1	0	0	0		
Tank 14	1 ± 1	0	0	0	0	0
PS 1	0	0	0	0	**0.4 ± 0.7** **0.17 ± 0.37**	**1.4 ± 2.0**
Jeddah PS	0	1 ± 1	0	3 ± 3		
Quiza PS	3	4 ± 1	1 ± 1	2 ± 2		
Mina A PS	4	6 ± 1	5 ± 1	3 ± 1		
Mina B PS	2	2 ± 2	0	2 ± 2		
Arafa Tank Outlet	8 ± 1	**1 ± 1**	0	0	**2.5 ± 1.8**	
Taif Tank Outlet	8 ± 1	**2 ± 2**	0	0		

Note that in stages 4 and 5 of the trial, a different set of parameters was measured, and this fact was not realized until the data sets were analyzed later.

The bromate results clearly show minimization of bromate in Tanks 3, 13, and 14, Pumping Station 1, and the Arafa and Taif tanks under all treatment conditions (Figure 4. The bromate values seen under condition 1 at the Arafa and Taif tanks reflect the time delay in transmitting water through the network; if one additional day is included in the control period, the Taif Tank values become 8 ± 2 for the control period and 0 for condition 1. This is the most significant result of this study: the demonstration that the procedure of sequential chlorination and ammoniation can effectively control bromate formation in transmitted water containing up to 300 ppb bromide which has been transported hundreds of kilometres at significant temperatures (average daytime maxima in Jeddah 34/37 °C April/May, in Taif 29/33 °C April/May). A bromate concentration below the target of 10 ppb was observed at Taif under the conditions prevailing in June/July (average daytime maxima Jeddah 38/39 °C, Taif 35/35 °C).

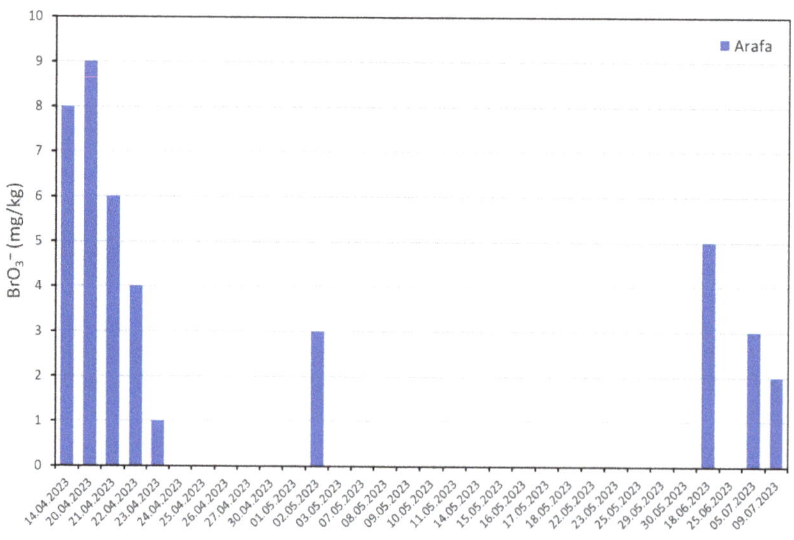

Figure 4. *Cont.*

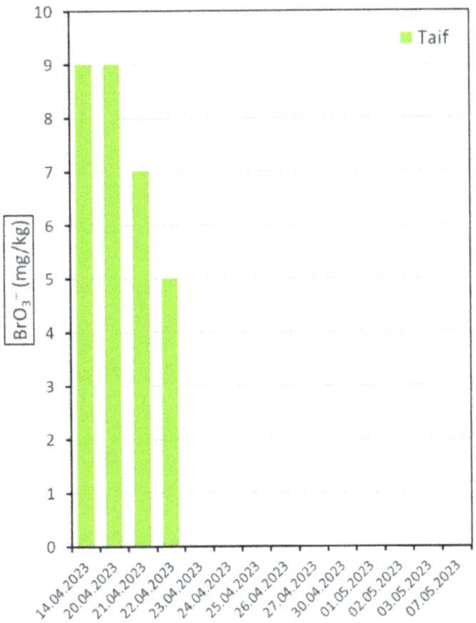

Figure 4. Bromate concentration in Arafa and Taif storage tanks during the 2023 trials.

The ammonium results obtained imply that significant amounts of ammonia are present in parts of the system that should not be affected by the trial, while at the same time low values are obtained at the Arafa and Taif tanks where excellent control of bromate was demonstrated (Table 8). Unfortunately, these results thus convey primarily concern about the methodology used to analyze for ammonium.

Table 8. Measured ammonium ion concentration (ppb) at locations within the Shoiabah desalination plant site and associated transmission system, 2023. Values **bold and underlined** are locations where water quality should be affected by ammoniation.

Stage	Control	1	2	3	4	5
Header–Phase 2					**121 ± 62**	**35 ± 59**
Phase 1–Tank 2	0	**92 ± 52**	**29 ± 30**			
Phase 1–Tank 3	0	**93 ± 32**	**74 ± 6**	**38 ± 52**	**21 ± 21**	**0**
Tank 4	0	**333 ± 58**	**337 ± 70**	**136 ± 136**		
Tank 13	0	**127 ± 76**	**150 ± 41**	**101 ± 59**		
Tank 14	0	**140 ± 77**	**156 ± 41**	**104 ± 36**	**86 ± 44**	**14 ± 23**
PS 1	0	**111 ± 46**	**44 ± 42**	**24 ± 24**	**46 ± 43** **404 ± 5**	**0**
Jeddah PS	0	77 ± 41	38 ± 32	0		
Quiza PS	0	103 ± 40	32 ± 18	0		
Mina A PS	0	51 ± 55	37 ± 31	0		
Mina B PS	0	78 ± 66	21 ± 28	49 ± 10		
Arafa Tank Outlet	100 ± 141	**150 ± 265**	**10 ± 12**	**10 ± 15**		
Taif Tank Outlet	0	**60 ± 7**	**30 ± 32**	**0**	0	

Increases in nitrate are seen across the system in the trial period compared to the control period, whether or not waters were subjected to chlorination/ammoniation treatment (Table 9). There is no consistent trend of a higher nitrate concentration with the addition of ammonia, but there is some evidence that ammonia addition may be leading to an increase in nitrate in that all tanks have a higher concentration of nitrate on first addition of ammonia in both studies—moving from the control case to condition 1 in this study, just as they did on moving from condition 0 to condition 1 in the 2022 study.

Table 9. Measured nitrate ion concentration (ppb) at locations within the Shoiabah desalination plant site and associated transmission system, 2023. Values **bold and underlined** are locations where water quality should be affected by ammoniation.

Stage	Control	1	2	3	4	5
Header–Phase 2					**8 ± 8**	**7 ± 12**
Phase 1–Tank 2	**5 ± 3**	**18 ± 15**	**40 ± 34**			
Phase 1–Tank 3	**5 ± 3**	**19 ± 8**	**27 ± 12**	**14 ± 11**	**37 ± 14**	**15 ± 15**
Tank 4	**18 ± 10**	**44 ± 17**	**57 ± 26**	**84 ± 7**		
Tank 13	**19 ± 6**	**23 ± 9**	**39 ± 28**	**68 ± 9**		
Tank 14	**11 ± 7**	**34 ± 14**	**30 ± 6**	**30 ± 6**	**19 ± 5**	**19 ± 28**
PS 1	**21**	**24 ± 12**	**31 ± 11**	**26 ± 6**	**19 ± 15** **22 ± 13**	**14 ± 22**
Jeddah PS	11	44 ± 13	34 ± 18	38 ± 14		
Quiza PS	15	20 ± 10	39 ± 20	74 ± 11		
Mina A PS	18	23 ± 11	29 ± 16	41 ± 23		
Mina B PS	8	16 ± 13	15 ± 7	9 ± 1		
Arafa Tank Outlet	14 ± 8	**17 ± 11**	**40 ± 11**	**67 ± 24**		
Taif Tank Outlet	11 ± 3	**34 ± 9**	**32 ± 14**	**29 ± 15**	**19 ± 24**	

It should also be noted that as part of the 2023 study, the waters at the Arafa and Taif tank outlets were tested for the presence of organic disinfection products on two dates: on May 25th, when ammoniation was carried out together with the addition of seawater to increase the bromide concentration to approximately 130 ppb, and on June 6th, when enough seawater to increase the bromide concentration to approximately 115 ppb was added. For the ammoniated sample, analysis was conducted for bromodichloromethane, bromoform, and dibromochloromethane, and in all cases concentrations were found to be below the detection threshold of 2 ppb. For the control sample, bromodichloromethane and dibromoacetonitrile were found to be below the detection threshold of 0.10 ppb, while bromoform was identified in the Arafa tank at a concentration of 1.92 ppb and the Taif tank at a concentration of 3.77 ppb, and dibromochloromethane was detected at a concentration of 0.27 ppb in the Arafa tank and 0.33 ppb in the Taif tank. While not as definitive as would be ideal, this indicates that disinfection of water containing 130 ppb bromide did not lead to detectable brominated organic by-product when combined with ammoniation. On June 6th, analysis was also done for monochloracetic acid (detection limit 1.0 ppb), dichloroacetic acid (detection limit 0.5 ppb) and trichloroacetic acid (detection limit 0.5 ppb), with none of these substances detected at either Arafa or Taif tanks, suggesting the treatment did not lead to an increase in haloacetic acids. Comprehensive tests were carried out by a third party for samples collected in August, when temperatures in Makkah province are at a maximum so the rate of generation of bromate and other disinfection by-products should be highest, and water at the Arafa and Taif tanks was found to meet regulatory limits for all disinfection by-products controlled by the Saudi Arabian authorities.

5. Conclusions

The combination of chlorination and ammoniation at levels of ammonia as low as 100 ppb has been demonstrated to effectively control the formation of bromate in water produced by seawater desalination on the commercial scale. Preliminary results suggest that the formation of brominate organic disinfection products was also controlled by this treatment. Consistent increases in the proportion of bromine measured as bromide were seen under the same conditions, suggesting that ammonia addition is at least in part controlling bromate formation by reducing the formation of hypobromite intermediate. Trends in nitrate concentration suggest that the addition of ammonia is not contributing significantly to the nitrate load in the product water. Most importantly, bromate control effects of ammonia addition were observed at water storage sites hundreds of kilometres from the seawater desalination plants under summer temperatures, suggesting that the course of treatment employed will control bromate formation in the transmission lines. This makes it a competitive approach to alternative capital-intensive solutions of second-stage RO to remove bromide at the source or post-treatment adsorption of bromate [49]. Further trials are ongoing within the SWCC network to explore chlorination–ammoniation as a cost-effective method for the control of brominated disinfection by-products.

Author Contributions: Conceptualization, A.A.A. (Ali A. Alhamzah); Methodology, A.A.A. (Ali A. Alhamzah), A.S.A. and A.A.A. (Abdulrahman A. Abid); Formal analysis, A.A.A. (Ali A. Alhamzah) and C.M.F.; Investigation, A.S.A. and A.A.A. (Abdulrahman A. Abid); Writing—original draft, C.M.F.; Writing—review & editing, A.A.A. (Ali A. Alhamzah); Project administration, A.A.A. (Ali A. Alhamzah). All authors have read and agreed to the published version of the manuscript.

Funding: This research received no external funding and was carried out as part of the authors' duties as employees of the SWCC.

Data Availability Statement: The data presented in this study are available on request from the corresponding author. The data are not publicly available due to SWCC policy.

Conflicts of Interest: Author Ali A. Alhamzah is named as an inventor on a patent application related to this work. The other authors declare no conflicts of interest.

References

1. Boorman, G.A. Drinking water disinfection byproducts: Review and approach to toxicity evaluation. *Environ. Health Perspect.* **1999**, *107*, 207–217. [CrossRef] [PubMed]
2. Li, X.-F.; Mitch, W.A. Drinking water disinfection byproducts (DBPs) and human health Effects: Multidisciplinary challenges and opportunities. *Environ. Sci. Technol.* **2018**, *52*, 1681–1689. [CrossRef] [PubMed]
3. Benítez, J.S.; Rodríguez, C.M.; Casas, A.F. Disinfection byproducts (DBPs) in drinking water supply systems: A systematic review. *Phys. Chem. Earth A/B/C* **2021**, *123*, 102987. [CrossRef]
4. Agus, E.; Sedlak, D.L. Formation and fate of chlorination by-products in reverse osmosis desalination systems. *Water Res.* **2010**, *44*, 1616–1626. [CrossRef]
5. Sharma, V.K.; Zboril, R.; McDonald, T.J. Formation and toxicity of brominated disinfection byproducts during chlorination and chloramination of water: A review. *J. Environ. Sci. Health B* **2014**, *49*, 212–228. [CrossRef]
6. Agus, E.; Voutchkov, N.; Sedlak, D.L. Disinfection by-products and their potential impact on the quality of water produced by desalination systems: A literature review. *Desalination* **2009**, *237*, 214–237. [CrossRef]
7. Antoniou, M.G.; Sichel, C.; Andre, K.; Andersen, H.R. Novel pre-treatments to control bromate formation during ozonation. *J. Hazard. Mater.* **2017**, *323*, 452–459. [CrossRef]
8. Bonacquisti, T.P. A drinking water utility's perspective on bromide, bromate, and ozonation. *Toxicology* **2006**, *221*, 145–148 [CrossRef] [PubMed]
9. Galey, C.; Gatel, D.; Amy, G.; Cavard, J. Comparative assessment of bromate control options. *Ozone Sci. Eng.* **2000**, *22*, 267–278. [CrossRef]
10. Gregov, M.; Jurinjak Tušek, A.; Valinger, D.; Benković, M.; Jurina, T.; Surać, L.; Kurajica, L.; Matošić, M.; Gajdoš Kljusurić, J.; Ujević Bošnjak, M.; et al. Linear and non-linear modelling of bromate formation during ozonation of surface water in drinking water production. *Water* **2023**, *15*, 1516. [CrossRef]
11. Ikehata, K.; Wang, L.; Nessl, M.B.; Komor, A.T.; Cooper, W.J.; McVicker, R.R. Effect of ammonia and chloramine pretreatment during the ozonation of a colored groundwater with elevated bromide. *Ozone Sci. Eng.* **2013**, *35*, 438–447. [CrossRef]
12. Ozekin, K.; Westerhoof, P.; Amy, G.L.; Siddiqui, M. Molecular ozone and radical pathways of bromate formation during ozonation. *J. Environ. Eng.* **1998**, *124*, 456–462. [CrossRef]
13. Pinkernell, U.; Von Gunten, U. Bromate minimization during ozonation: Mechanistic considerations. *Environ. Sci. Technol.* **2001**, *35*, 2525–2531. [CrossRef]
14. Soltermann, F.; Abegglen, C.; Tschui, M.; Stahel, S.; Von Gunten, U. Options and limitations for bromate control during ozonation of wastewater. *Water Res.* **2017**, *116*, 76–85. [CrossRef] [PubMed]
15. Von Gunten, U. Ozonation of drinking water: Part II. Disinfection and by-product formation in presence of bromide, iodide or chlorine. *Water Res.* **2003**, *37*, 1469–1487. [CrossRef] [PubMed]
16. Ministry of Environment, Water and Agriculture. *Standards and Specifications for Water Types*; Ministry of Environment, Water and Agriculture: Riyadh, Saudi Arabia, 1442.
17. Ministry of Environment, Water and Agriculture. *Guide to Monitoring Unbottled Drinking Water Quality Standards in the Water Service Delivery Chain*; Ministry of Environment, Water and Agriculture: Riyadh, Saudi Arabia, 1444.
18. Morris, J.C. The acid ionization constant of HOCl from 5 to 35°. *J. Phys. Chem.* **1966**, *70*, 3798–3805. [CrossRef]
19. Achmit, M.; Machkor, M.; Nawdali, M.; Sbai, G.; Karim, S.; Aouniti, A.; Loukili, M. Study of the influence of the operating parameters on the fractions in HOCl and OCl^- during the disinfection phase. *J. Chem. Pharm. Res.* **2018**, *10*, 122–127.

20. Kumar, K.; Margerum, D.W. Kinetics and mechanism of general-acid-assisted oxidation of bromide by hypochlorite and hypochlorous acid. *Inorg. Chem.* **1987**, *26*, 2706–2711. [CrossRef]
21. Beckwith, R.C.; Margerum, D.W. Kinetics of hypobromous acid disproportionation. *Inorg. Chem.* **1997**, *36*, 3754–3760. [CrossRef]
22. Al-Hamzah, A.A.; Mahmoodur Rahman, M.; Kurup, P.; Barnawi, A.; Ghannam, B.; Musharraf, I.; Al-Najjar, F.; Obeidallah, A.; Palmer, N. Use of chlorine dioxide as alternative to chlorination in reverse osmosis product water. *Desalin. Water Treat.* **2019**, *163*, 57–66. [CrossRef]
23. Margerum, D.W.; Huff Hartz, K.E. Role of halogen(I) cation-transfer mechanisms in water chlorination in the presence of bromide ion. *J. Environ. Monit.* **2002**, *4*, 20–26. [CrossRef] [PubMed]
24. Ling, L.; Deng, Z.; Fang, J.; Shang, C. Bromate control during ozonation by ammonia-chlorine and chlorine-ammonia pretreatment: Roles of bromine-containing haloamines. *Chem. Eng. J.* **2020**, *389*, 123447. [CrossRef]
25. Bratsch, S.G. Standard electrode potentials and temperature coefficients in water at 298.15 K. *J. Phys. Chem. Ref. Data* **1989**, *18*, 1–21. [CrossRef]
26. Jafvert, C.T.; Valentine, R.L. Reaction scheme for the chlorination of ammoniacal water. *Environ. Sci. Technol.* **1992**, *26*, 577–586. [CrossRef]
27. Wajon, J.E.; Morris, J.C. Rates of formation of N-bromo amines in aqueous solution. *Inorg. Chem.* **1982**, *21*, 4258–4263. [CrossRef]
28. Qiang, Z.; Adams, C.D. Determination ofmonochloramine formation rateconstants with stopped-flow spectrophotometry. *Environ. Sci. Technol.* **2004**, *38*, 1435–1444. [CrossRef]
29. Sugam, R.; Helz, G.R. Chlorine speciation in seawater; a metastable equilibrium model for ClI and BrI species. *Chemosphere* **1981**, *10*, 41–57. [CrossRef]
30. Gazda, M.; Margerum, D.W. Reactions of Monochloramine with Br_2, Br_3, HOBr, OBr^-: Formation of Bromochloramines. *Inorg. Chem.* **1994**, *33*, 118–123. [CrossRef]
31. Lei, H.; Mariñas, B.J.; Minear, R.A. Bromamine decomposition kinetics in aqueous solutions. *Environ. Sci. Technol.* **2004**, *38*, 2111–2119. [CrossRef]
32. Gazda, M.; Dejarme, L.E.; Choudhury, T.K.; Cooks, R.G.; Margerum, D.W. Mass-spectrometric evidence for the formation of bromochloramine and N-bromo-N-chloromethylamine in aqueous solution. *Environ. Sci. Technol.* **1993**, *27*, 557–561. [CrossRef]
33. Trofe, T.W.; Inman, G.W.; Johnson, J.D. Kinetics of monochloramine decomposition in the presence of bromide. *Environ. Sci. Technol.* **1980**, *14*, 544–549. [CrossRef]
34. Heeb, M.B.; Criquet, J.; Zimmermann-Steffens, S.G.; Von Gunten, U. Oxidative treatment of bromide-containing waters: Formation of bromine and its reactions with inorganic and organic compounds—A critical review. *Water Res.* **2014**, *48*, 15–42. [CrossRef] [PubMed]
35. Luh, J.; Mariñas, B.J. Kinetics of bromochloramine formation and decomposition. *Environ. Sci. Technol.* **2014**, *48*, 2843–2852. [CrossRef]
36. Zhu, X.; Zhang, X. Modeling the formation of TOCl, TOBr and TOI during chlor(am)ination of drinking water. *Water Res.* **2016**, *96*, 166–176. [CrossRef] [PubMed]
37. Brodfuehrer, S.H.; Goodman, J.B.; Wahman, D.G.; Speiteland, G.E.; Katz, L.E. Apparent reactivity of bromine in bromochloramine depends on synthesis method: Implicating bromine chloride and molecular bromine as important bromine species. *J. Environ. Eng.* **2022**, *148*, 06022006. [CrossRef]
38. Valentine, R.L. Bromochloramine oxidation of N,N-diethyl-p-phenylenediamine in the presence of monochloramine. *Environ. Sci. Technol.* **1986**, *20*, 166–170. [CrossRef]
39. Gilbert, R.G.; (University of Sydney, Sydney, Australia). Personal communication, 1999.
40. Trogolo, D.; Arey, J.S. Equilibria and speciation of chloramines, bromamines, and bromochloramines in water. *Environ. Sci. Technol.* **2017**, *51*, 128–140. [CrossRef] [PubMed]
41. Pope, P.G.; Speitel, G.E. Reactivity of bromine-substituted haloamines in forming haloacetic acids. In *Disinfection By-Products in Drinking Water*; Karanfil, T., Krasner, S.W., Westerhoff, P., Xie, Y., Eds.; American Chemical Society: Washington, DC, USA, 2008; pp. 182–197. [CrossRef]
42. Simon, V.; Berne, F.; Gallard, H. Chloramination and bromamination of amino acids. In *Disinfection By-Products in Drinking Water*; Thompson, K.C., Gillespie, S., Goslan, E., Eds.; The Royal Society of Chemistry: Cambridge, UK, 2015; pp. 70–80. [CrossRef]
43. Alsulaili, A.; Speitel, G.E.; Katz, L.E. Monochloramine and total haloamine decay after a short prechlorination time in the presence of bromide. *Water Supply* **2010**, *10*, 512–516. [CrossRef]
44. Sun, Y.-X.; Wu, Q.-Y.; Hu, H.-Y.; Tian, J. Effect of ammonia on the formation of THMs and HAAs in secondary effluent chlorination. *Chemosphere* **2009**, *76*, 631–637. [CrossRef]
45. Hu, W.; Croué, J.-P.; Allard, S. Effect of copper oxide on monochloramine decomposition in bromide-containing waters. *Sci. Total Environ.* **2021**, *765*, 142519. [CrossRef]
46. Standard Methods Committee of the American Public Health Association; American Water Works Association; Water Environment Federation. 4500-H+ pH. In *Standard Methods for the Examination of Water and Wastewater*, 23rd ed.; Lipps, W.C., Baxter, T.E., Braun-Howland, E., Eds.; APHA Press: Washington, DC, USA, 2017.
47. Standard Methods Committee of the American Public Health Association; American Water Works Association; Water Environment Federation. 4110 Determination of anions by ion chromatography. In *Standard Methods for the Examination of Water and Wastewater*, 23rd ed.; Lipps, W.C., Baxter, T.E., Braun-Howland, E., Eds.; APHA Press: Washington, DC, USA, 2017.

48. Pearce, R.; Hogard, S.; Buehlmann, P.; Salazar-Benites, G.; Wilson, C.; Bott, C. Evaluation of preformed monochloramine for bromate control in ozonation for potable reuse. *Water Res.* **2022**, *211*, 118049. [CrossRef]
49. Morrison, C.M.; Hogard, S.; Pearce, R.; Mohan, A.; Pisarenko, A.N.; Dickenson, E.R.V.; Von Gunten, U.; Wert, E.C. Critical review on bromate formation during ozonation and control options for its minimization. *Environ. Sci. Technol* **2023**. *online ahead of print*. [CrossRef]

water MDPI

Article

Cyanide Bioremediation by *Bacillus subtilis* under Alkaline Conditions

César Julio Cáceda Quiroz [1,*], Gabriela de Lourdes Fora Quispe [1,*], Milena Carpio Mamani [1], Gisela July Maraza Choque [1] and Elisban Juani Sacari Sacari [2]

[1] Laboratorio de Biorremediación, Facultad de Ciencias, Universidad Nacional Jorge Basadre Grohmann, Avenida Miraflores S/N, Ciudad Universitaria, Tacna 23003, Peru

[2] Centro de Energías Renovables de Tacna (CERT), Facultad de Ciencias, Universidad Nacional Jorge Basadre Grohmann, Avenida Miraflores S/N, Ciudad Universitaria, Tacna 23003, Peru

* Correspondence: ccacedaq@unjbg.edu.pe (C.J.C.Q.); gforaq@unjbg.edu.pe (G.d.L.F.Q.)

Abstract: Cyanide (CN) is a toxic environmental pollutant generated by various industrial activities, necessitating the application of bioremediation techniques for its degradation. Biodegradation is a more cost-effective and environmentally friendly technique with high efficiency in CN removal. This study isolated cyanide-degrading bacteria from Tutupaca mining site soil from Tacna, Peru. *Bacillus subtilis* strain TT10s was selected for its exceptional capacity to rapidly and completely eliminate cyanide under alkaline conditions (pH 10.5), removing 1000 ppm cyanide within 48 h. A kinetic analysis revealed that the biodegradation follows second-order rate kinetics (k_2 = 0.08649 mg/(mg·h), R^2 = 0.96622), consistent with the literature attribution of the rate-limiting step to the inducible cyanide dihydratase enzyme, which converts cyanide into ammonia and formate via the Michaelis–Menten model. Fourier-transform infrared spectroscopy (FTIR) spectral analysis further corroborated this enzymatic mechanism, showing the disappearance of CN peaks coupled with the emergence of ammonia (NH) and formate (C=O) peaks. Quantitative kinetic modelling integrated with FTIR profiles and degradation curves implicates cyanide dihydratase as the key rate-controlling enzyme in alkaline cyanide biodegradation without the need for an extra carbon source, generating interest for future bioremediation applications in highly contaminated environments.

Keywords: bioremediation; cyanide; *Basillus subtilis*

1. Introduction

Cyanide is a highly toxic pollutant generated from industrial activities like electroplating [1], mining [2], the production of organic chemicals [3], and others.

Cyanide is extensively used in mining, and Peru is not immune to the environmental impacts of cyanide [4] due to it being extensively used for gold and copper extraction [5]. This has resulted in severe environmental impacts from cyanide spills [4], leaks [6], and the improper discharge of contaminated wastewater into rivers and lakes, which has been exacerbated by the Peruvian government declaring states of emergency in certain regions [7]. Conflicts have emerged over water quality and supply between mining companies and local communities throughout Peru, especially in agricultural regions dependent on clean water [8,9]. Peru has developed regulations on cyanide use and discharge limits [10]; however, the strict enforcement of these regulations is necessary to ensure their efficacy, as stakeholders may ignore them.

Studies have detected high levels of cyanide contamination in different rivers in Tumbes [11], Tacna [12,13], Amazonas [14], Piura [15], Arequipa [16,17], and other regions in Peru, likely originating from legal and illegal mining operations [18].

The Tacna region in Peru faces several environmental liabilities left behind by abandoned informal mines where various materials, including cyanide, were used for gold, copper, and sulfur processing, reaching concentrations of up to 92 mg/kg of free cyanide in the Cano community [13]. These sites are located close to rivers, leading to risks of water contamination that could have negative effects on the environment and health of people consuming the polluted water [13].

The accumulation of cyanide in the environment poses risks to ecosystems and human health due to its ability to inhibit essential metalloenzymes [19,20]. Conventional physicochemical methods for cyanide removal have limitations like high costs, secondary pollution, and the inability to completely destroy cyanide [21]. Bioremediation techniques using microbes provide a promising, eco-friendly solution to decontaminate cyanide-polluted sites and prevent further environmental damage [22,23]. A range of bacterial species have been studied for their ability to degrade cyanide, either as the sole source of nitrogen or co-metabolically with other substrates. Common cyanide-degrading bacteria include species from the genera *Pseudomonas, Bacillus, Klebsiella, Burkholderia,* and *Rhodococcus.* Within the *Pseudomonas genus, P. pseudoalcaligenes* [24], *P. fluorescens* [25], and *P. putida* [26] display cyanide removal capabilities. *Bacillus megaterium* [27], *B. subtilis* [28], *B. pumilus* [29], and *B cereus* [30] strains have also been investigated. Other promising cyanide-degrading bacteria include *Klebsiella oxytoca* [31], *Burkholderia cepacia* [32], and *Rhodoccocus sp.* [33]. These bacteria possess enzymes like cyanide dioxygenase, cyanidase, nitrilase, and rhodanese that allow them to transform cyanide into less toxic byproducts [34].

Bacillus subtilis, a ubiquitous Gram-positive bacterium, has been extensively studied for its cyanide removal abilities, making it a promising candidate for bioremediation [28,35–37]. Under alkaline conditions (pH 9–10), *B. subtilis* can efficiently degrade high cyanide concentrations (900 ppm) by expressing cyanide dihydratase enzymes [30].

Alkaline conditions favor cyanide biodegradation by reducing the volatilization of toxic HCN gas, enabling complete cyanide mineralization [34]. The high pH also provides selective advantages to alkaliphilic *B. subtilis* to outcompete other microbes. Nutrient supplementation further enhances cyanide transformation by stimulating bacterial growth [38]. Immobilized cells and genetically engineered strains of *B. subtilis* with increased cyanide dihydratase expression can be applied for efficient cyanide remediation [39].

The use of bacteria offer advantages for cyanide bioremediation, such as rapid growth, the ease of genetic modification, and an ability to withstand harsh conditions [40]. They can degrade high cyanide concentrations, operate over a range of pH levels and temperatures, and require simple nutritional inputs [30]. Applications as suspended cultures or immobilized cells enhance the stability and cyanide degradation capacity [41]. Additionally, engineered and alkaliphilic bacterial strains can be designed for optimized biodegradation under specific conditions [42].

In this study, we isolated and identified a native bacterial consortium from a sulfur mining environmental liability in Tacna, Peru. The consortium was enriched from tailing soil samples under alkaline conditions (pH 10.5) to select for bacteria adapted to the high pH, and relevant higher cyanide concentration parameters for cyanide-containing mining effluents. Batch biodegradation experiments were conducted to evaluate the cyanide removal kinetics and the efficiency of the selected bacterial strain at an alkaline pH using initial cyanide concentrations and conditions representative of real mining wastewaters.

Fourier-transform infrared (FTIR) spectroscopy was used to study the functional groups of compounds made during the biodegradation of alkaline cyanide to figure out the specific biochemical mechanisms and pathways involved.

2. Materials and Methods

2.1. Sample Collection

Soil samples were collected in high density polypropylene bags from the Tutupaca (east 0358602; north 8113940; altitude 4687 m.s.n.m.) mining environmental liabilities located in the Candarave province of Tacna, Peru. The soil samples were then homogenized and sieved to obtain 100 g of each soil sample. These were cultured in 500 mL Erlenmeyer flasks containing 100 mL of sterile distilled water, which was homogenized for 15 min to obtain the supernatant. To enrich and isolate alkaliphilic bacteria, 10 mL of the supernatant was diluted in 90 mL of sterile nutrient broth (Merck Company, Darmstadt, Germany) and incubated at 30 °C for 24 h at 150 rpm. Then, 10 mL of this culture was inoculated into 90 mL of nutrient broth with the pH adjusted to 10.5 using sterile 0.1 N NaOH, and incubated at 30 °C for 24 h at 150 rpm. This alkaline-adapted culture was then streak-plated on nutrient agar plates (Merck Company, Darmstadt, Germany) to isolate and purify bacterial colonies capable of growing at pH 10.5 [30]. The obtained samples were labeled as TT1s, TT3s, TT7s, TT8s, TT9s, TT10s, TT11s, and TT13s, respectively.

2.2. Bacterial Isolation

To detect cyanide-degrading bacteria, 10 mL of the enriched culture was transferred to a 250 mL Erlenmeyer flask containing 100 mL of M9 minimal medium (g/L): $Na_2HPO_4 \cdot 7H_2O$ (12.8); KH_2PO_4 (3); NaCl (0.5); $MgSO_4 \cdot 7H_2O$ (0.5); $CaCl_2$ (0.1); 0.2% (*w*/*v*) sodium acetate; 0.2% (*w*/*v*) yeast extract; and 1% (*v*/*v*) mineral salt solution containing (g/L): $ZnSO_4 \cdot 7H_2O$ (0.05); $MnCl_2 \cdot 4H_2O$ (0.05); $CuCl_2 \cdot 2H_2O$ (0.005); $Na_2MoO_4 \cdot 2H_2O$ (0.005); $Na_2B_4O_7 \cdot 10H_2O$ (0.002); and $CoCl_2 \cdot 6H_2O$ (0.0003). This was supplemented with NaCN concentrations from 100 to 1000 ppm, with the pH adjusted to 10.5 using sterile NaOH. After incubation, the viability of the isolated bacteria at each NaCN concentration was verified by streak-plating on nutrient agar plates. An alkaline pH above 10.5 was maintained to minimize the volatilization of cyanide as HCN [43–45]. Bacteria viable at cyanide levels above 1000 ppm were selected for further characterization based on their cyanide degradation potential. Gramme staining and biochemical testing were performed to verify the microscopic characteristics and taxonomy of the isolates [44].

2.3. Sequencing and Molecular Identification

To perform the high-throughput sequencing of bacterial genomes, high molecular weight genomic DNA was first extracted from pure cultures grown on Luria Bertani agar for 24 h at 35 °C. The InnuPREP Bacteria DNA Kit (Analytik Jena, Jena, Germany) was used for extraction, which employs enzymatic, mechanical, and chemical lysis for robust cell disruption followed by selective DNA binding to a silica spin filter for purification. DNA concentrations were then accurately quantified through fluorometry using a Qubit 4 fluorometer (Life Technologies, Carlsbad, CA, USA). For sequencing library preparation, 100 ng of genomic DNA was fragmented, end-repaired, and ligated with Illumina adapters using the Illumina DNA Prep workflow (Illumina, Granta Park, UK). Unique dual indexes from the Nextera DNA CD Indexes kit (Illumina, Granta Park, UK) were added to libraries during PCR enrichment to enable the multiplexing of samples. Finally, libraries were sequenced as 2 × 151 bp paired-end reads on the Illumina Miseq platform using a 600-cycle reagent kit. The 16S rRNA gene sequence obtained from the potential cyanide-degrading bacterial isolate was analyzed via a comparative identity search using the BLAST (Basic Local Alignment Search Tool) against the National Center for Biotechnology Information (NCBI) database.

2.4. Bacterial Growth Kinetics

Bacterial growth kinetics were quantified by optical density measurements and viable plate counting. Optical density at 600 nm (OD600) was measured using a spectrophotometer (Epoch 2c, Biotek, Winooski, VT, USA) to generate high-throughput growth curves showing population expansion over time. This provided an indirect estimate of the overall cell density in the culture [45]. However, OD600 does not discriminate between viable and dead cells. Therefore, viable plate counting was also carried out through dilution plating on agar and colony formation, allowing the direct quantification of colony-forming units (CFU).

2.5. Cyanide Biodegradation

To study the cyanide degradation capability of the selected alkaliphilic bacterial isolates, 150 mL of the M9 minimal medium was prepared and supplemented with 1000 ppm sodium cyanide. This was inoculated with the bacteria at a density of 10^7–10^8 cells/mL and incubated at 35 °C for 48 h with aeration at 0.2 VVM. Aliquots were taken at specific time intervals in triplicate to quantify the remaining cyanide concentration over time. The concentration of free cyanide in aliquots was determined using a titrimetric method based on Standard Methods 4500-CN-D [46]. This method relies on the reaction between silver nitrate ($AgNO_3$) and cyanide ions (CN^-) in alkaline solution to form the soluble complex silver cyanide (Ag(CN)). Specifically, a 10 mL aliquot of the sample was analyzed by adding 3 drops of potassium iodide (KI), which acts as an indicator. The sample was then titrated with a standardized $AgNO_3$ solution. The Ag^+ ions reacted with CN^- until all free cyanide had been consumed, at which point excess Ag^+ reacted with the KI^- indicator to form a precipitate, marking the endpoint. The amount of $AgNO_3$ required to reach this endpoint allows the quantification of the original free cyanide concentration. A ratio proposed by Copari et al. [12] was used to calculate the concentration, where 1 mL of $AgNO_3$ titrant corresponds to 20 ppm of free cyanide (CN^-) in the sample.

The cyanide degradation efficiency (*DE*) of the bacterial strain was calculated using the following formula [47]:

$$DE = \frac{I_c - R_c}{I_c} \cdot 100$$

where I_c is the initial concentration of cyanide (mg/L) and R_c is the residual concentration of cyanide (mg/L) after treatment.

2.6. Kinetic Biodegradation Models

To analyze the kinetics of cyanide biodegradation by the microorganisms, first- and second-order rate models were fit to the experimental data [48]. The first-order model assumes the reaction rate is proportional to the cyanide concentration, described mathematically as

$$\frac{dS}{dt} = k_1 S$$

where S is the cyanide concentration (mg/L), t is time (h), and k_1 (1/h) is the first-order rate constant. The integration of this differential rate law gives

$$S = S_0 e^{-k_1 t}$$

where S_0 is the initial cyanide concentration (mg/L).

The second-order model assumes a more complex kinetic rate law proportional to both cyanide concentration and biomass concentration:

$$\frac{dS}{dt} = -k_2 S X$$

where X is biomass concentration (mg/L) and k_2 is the second-order rate constant (mg/(mg·d)). This integrates to

$$k_s ln\left(\frac{S}{S_0}\right) + S - S_0 = -k_2 t$$

where k_s is the half-saturation coefficient (mg/L).

These kinetic models were fit to experimental cyanide degradation data to estimate rate constants. The model providing the best fit, as determined by the highest correlation coefficient (R^2), suggests the order of reaction and rate-limiting steps.

2.7. Analytical Methods

To study the degradation and transformation of cyanide over time, samples were collected at 0, 24, and 48 h. The samples were centrifuged at 8000 rpm for one minute to separate the cells from the supernatant. The cell-free supernatant was then analyzed using a

Fourier-transform infrared spectrometer (FTIR) equipped with an attenuated total reflection (ATR) accessory (Bruker Optics, Invenio R, Ettlingen, Germany). FTIR-ATR provides a rapid, non-destructive method to characterize molecular changes in the supernatant, letting us investigate the metabolic conversion of cyanide and identify intermediates and products formed during biodegradation. For each sample, spectra were recorded from 4000 to 400 cm^{-1} at a resolution of 2 cm^{-1}. Twenty scans were co-added to improve the signal-to-noise ratio, and the average spectrum was used for analysis.

3. Results

3.1. Bacterial Isolation and Selection

Thirteen bacterial strains were isolated from soil samples collected at the Tutupaca mining site in Perú, which represents an environmental liability due to legacy contamination. Of these isolates, strains TT1s, TT3s, TT7s, TT8s, TT9s, TT10s, TT11s, and TT13s demonstrated growth viability in the presence of 100 ppm, 50 ppm, 300 ppm, 300 ppm, 600 ppm, 1000 ppm, 1000 ppm, and 600 ppm cyanide, respectively, at an alkaline pH of 10.5 (Figure 1). The isolates included Gram-negative and Gram-positive rods and cocci, with a predominance of Gram-positive, spore-forming bacilli (Table 1).

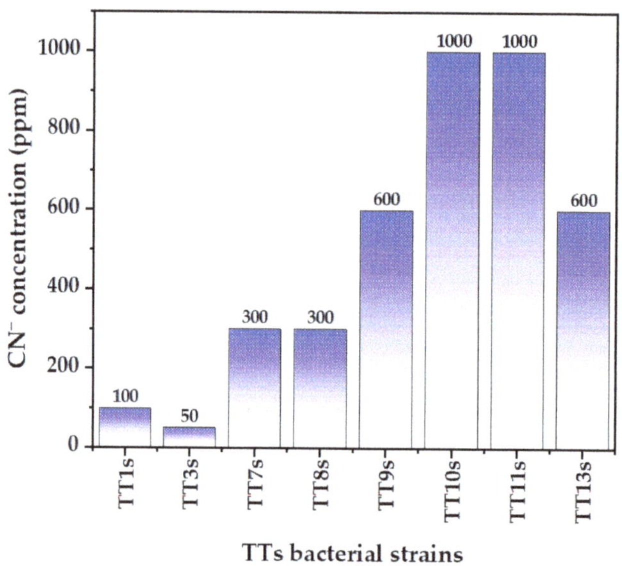

Figure 1. Bacterial strains surviving Tutupaca mining conditions: adapting to different cyanide Concentrations at pH 10.5 in 9 M minimal mineral medium.

Table 1. Morphological features of CN^--resilient bacterial strains.

Bacterial Code	Gram Reaction	Cell Morphology	Spore Formation	Motility
TTs1	Positive	Rod chain	+	+
TTs3	Negative	Rods chain	−	−
TT7s	Negative	Rod	−	+
TT8s	Positive	Coco	−	−
TT9s	Positive	Rods	+	+
TT10s	Positive	Rods	+	+
TT11s	Positive	Rods chain	+	+
TT13s	Negative	Rods	−	+

The ability of these native isolates to tolerate high cyanide levels under alkaline conditions reflects adaptations to the selective pressure imposed by legacy mining pollution at the site [49]. The predominance of Bacillus species is consistent with prior research showing their competence in alkaline cyanide biodegradation [50–52]. The cyanide tolerance of these isolates suggests they likely possess enzymatic pathways for cyanide transformation as a detoxification mechanism [53].

The various biochemical assays were performed for the selection strain TT10s, due to its capacity for viable survival up to 1000 ppm and its quick growth. The results coincided with the genus Bacillus (Table 2) [54]. Specifically, TT10s was found to be catalase and oxidase positive, methyl red negative, Voges–Proskauer positive, urease negative, able to use citrate, and capable of nitrate reduction. The strain could utilize glucose, mannitol, lactose, D-xylose, and sucrose as sole carbon sources, but not maltose [55]. No indole production was detected. Taken together, these biochemical characteristics are consistent with *Bacillus* spp. [56], supporting the identification of TT10s as a novel Bacillus strain with potential industrial applications due to its high tolerance to toxins.

Table 2. Biochemical analysis of CN^--resilient bacterial strain TT10s.

Bacterial	**Catalase**	**Oxidase**	**MR**	**VP**	**Citrate**	**Urease**	**Nitrate**	**Glucose**	**Maltose**	**Mannitol**	**Lactose/N_2**	**D-Xylose**	**Sucrose/ Indole**
TT10s	+	+	−	+	+	−	+	+	−	+	+	+	+

3.2. Molecular Identification

The 16S rRNA gene sequence of the isolated strain TT10s was elucidated using the BLAST alignment tool and compared to nucleotide sequences in the NCBI GenBank database. The nearly complete 1538 bp sequence exhibited 100% similarity with three Bacillus species: *Bacillus rugosus* (sequence IDs: MT554518.1, NR_181236.1, CP096590.1), *Bacillus subtilis* (sequence IDs: CP 025941.1, CP 018172.1, KR967391.1, OL636042.1), and *Bacillus stercoris* (sequence IDs: CP126678.1, CP124601.1). A phylogenetic analysis using the maximum likelihood algorithm positioned TT10s within the *B. subtilis* clade with strong bootstrap support. Taken together, this molecular identification aligned with the observed morphological properties of TT10s, including Gram-positive rods that form endospores and biochemical characteristics such as catalase and urease activity and the ability to utilize D-xylose and citrate (Tables 1 and 2). Based on an integrative taxonomic analysis of phenotypic, microscopic, and 16S rRNA sequence data, the isolated bacterial strain TT10s can be definitively classified as belonging to *Bacillus subtilis* because this species is urease positive and D-xylose negative, in contrast to *Bacillus stercoris* and *Bacillus rugosus*, which are negative and positive, respectively [57–59]. The gene sequence was deposited in the GenBank nucleotide archive under accession number OR505001 for reference.

3.3. Growth Kinetics

The relation between the OD optical density and the natural logarithm of the bacterial growth in CFU/mL of the TT10s bacterium was determined by means of a linear adjustment of its growth curve (Figure 2), determining an R^2 of 0.94 for the TT10s bacterium. Although this method may interfere with the results, due to the detection of non-cellular solids in the samples, it is advantageous due to its speed and automation [60]. For this reason, this calibration curve was made with replicas to estimate the bacterial growth of live cells in the biodegradation process of the TT10s bacterium.

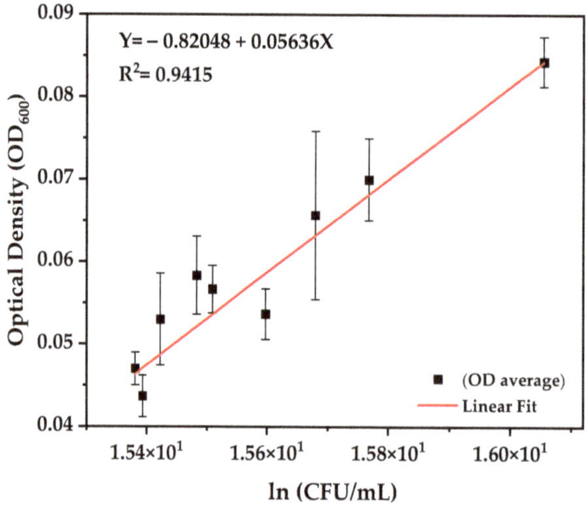

Figure 2. Linear fitting of bacterial growth for bacterial strain TT10s in OD600 (Abs) and Ln plate count (CFU/mL).

3.4. Cyanide Biodegradation

The biodegradation study of free cyanide was carried out via sampling every 6 h for the bacterial strain TT10s, as it was the strain with the greatest viability capacity at concentrations up to 1000 ppm in the M9 mineral medium at a pH of 10.5, temperature of 35 °C, and 0.2 vvm. This TT10s bacterial strain presented a degradation of 1000 ppm cyanide in less than 48 h (Table 3), with a degradation efficiency of up to 100% at 48 h and a maximum bacterial growth of up to 1.583×10^1 (ln CFU/mL) at 18 h (Figure 3).

Table 3. Cyanide degradation efficiency (%) and Ln bacterial growth (CFU/mL) for TT10s bacteria.

Time (h)	CN^- Degradation Efficiency (%)	SD	Plate Count Ln (CFU/mL)	SD
0	0.00	0.00	15.70	0.00
6	20.67	1.53	15.55	0.04
12	41.07	1.10	15.64	0.09
18	63.60	0.40	15.83	0.06
24	86.87	2.58	15.58	0.03
30	90.80	2.23	15.51	0.05
36	94.53	0.46	15.38	0.05
42	99.63	0.06	15.26	0.01
48	100.00	0.00	15.18	0.01

Bacterial growth was measured with ln CFU/mL, and peaked at 1.583×10^1 at 18 h before decreasing. This growth curve is typical for batch culture, with exponential growth followed by stationary and death phases [30].

The rapid and complete cyanide biodegradation by TT10s demonstrates its potential for industrial applications in treating cyanide waste streams compared to other bacterial strains, according to Table 4. The alkaline conditions prevented volatile HCN formation, enhancing process safety [61].

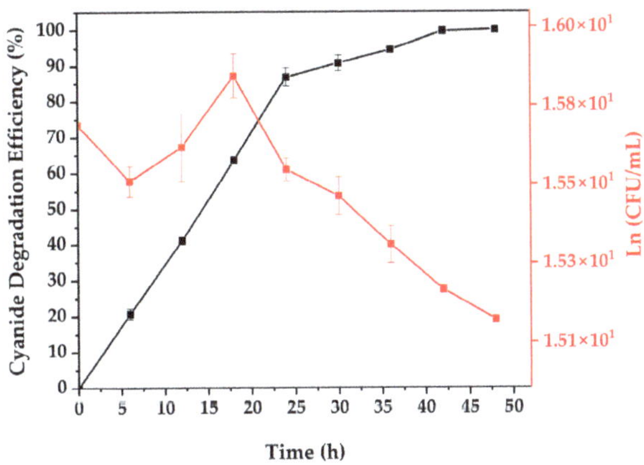

Figure 3. Cyanide degradation efficiency% in relation to time and *Bacillus subtilis* bacterial growth within 48 h.

Table 4. Cyanide biodegradation performances of various bacteria.

Bacterium	Experimental Conditions	Cyanide Degradation	Reference
Bacillus subtilis	48 h, pH 10.5, 30 °C, 1000 mg/L initial cyanide	100%	Present work
Pseudomonas putida	24 h, pH 9, 30 °C, 100 mg/L initial cyanide	90%	[26]
Pseudomonas fluorescens	48 h, pH 7, 30 °C, 50 mg/L initial cyanide	80%	[62]
Bacillus sp.	96 h, pH 9.88, 33.6 °C, 500 mg/L initial cyanide	99%	[51]
Klebsiella pneumoniae	72 h, pH 7, 25 °C, 25 mg/L initial cyanide	87%	[63]
Bacillus cereus	48 h, pH 7, 37 °C, 100 mg/L initial cyanide	95.87%	[64]

The isolated *Bacillus subtilis* strain TT10s demonstrated exceptionally rapid and complete cyanide degradation under alkaline conditions compared to the other bacteria studied. As shown in Table 4, TT10s achieved a 100% removal of a high initial cyanide concentration of 1000 mg/L within 48 h at pH 10.5, significantly outperforming various Pseudomonas, *Klebsiella*, and other Bacillus species strains, which required longer times or achieved only 60–90% removal at lower cyanide loadings and pH. The kinetic analysis indicates the specialized enzymatic pathways of TT10s enabled uniquely efficient cyanide metabolism rates even at elevated pH where volatile toxic HCN production is minimized. The rapid total degradation by TT10s shows great promise for the industrial bioremediation of alkaline cyanide effluents compared to currently studied bacteria.

3.5. Kinetic Models

Kinetic modeling was performed to elucidate the rate law and mechanisms governing cyanide biodegradation by *Bacillus subtilis*. Determining the most appropriate model that accurately depicts the kinetics is crucial for mechanistic insight. The experimental data strongly fit a second-order rate model (k_2 = 0.08649 mg/(mg·h), R^2 = 0.96622) better than a first-order model (k_1 = 0.11394 1/h, R^2 = 0.8514), as shown in Figure 4. Several factors justify selecting the second-order kinetics, but principally they exhibited a statistically stronger empirical fit with a higher coefficient of determination (R^2), agreeing with the literature in showing second-order dependence on the rate-limiting cyanide dihydratase enzyme [26,34,65]. Cyanide dihydratase catalyzes the conversion of cyanide to ammonia and formate via Michaelis–Menten kinetics as an inducible enzyme requiring cyanide as a substrate [65–68]. The higher coefficient of determination (R^2) and lower second-order rate constant (k_2) compared to the first-order rate constant (k_1) indicate that the second-order model more accurately reflects the enzymatic rate-limiting mechanism. Additional evidence comes from the observed lag phase, of up to 12 h, attributed to enzyme induction [69], an exponential phase from 12 to 30 h as enzyme levels increase, and a plateau after 30 h, indicating saturation kinetics.

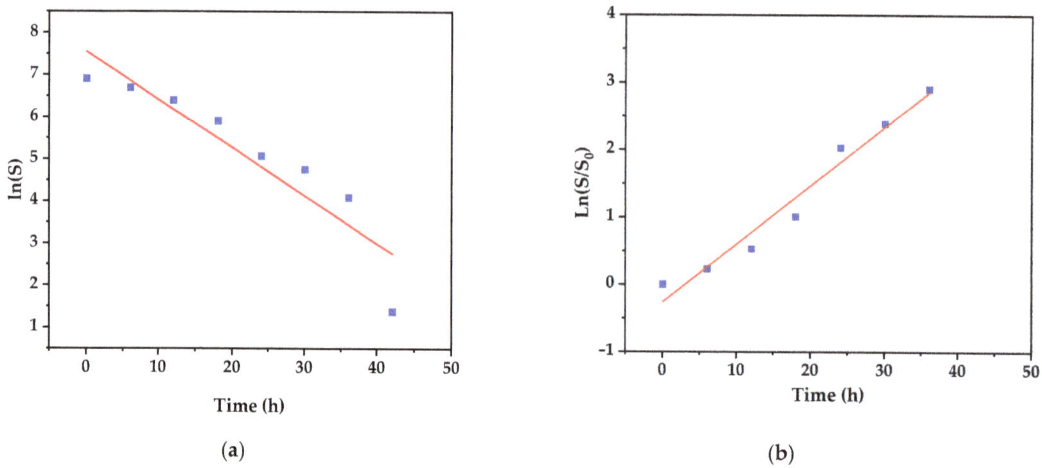

Figure 4. Kinetic model of the cyanide degradation; (**a**) first order and (**b**) second order fitting plots.

3.6. FTIR Spectroscopy Analysis

The FTIR spectra shown in Figure 5 indicate several key changes occurring during the 48 h period that align with cyanide being degraded into ammonia and formate by the *Bacillus subtilis* strain via the cyanide dihydratase enzymatic pathway.

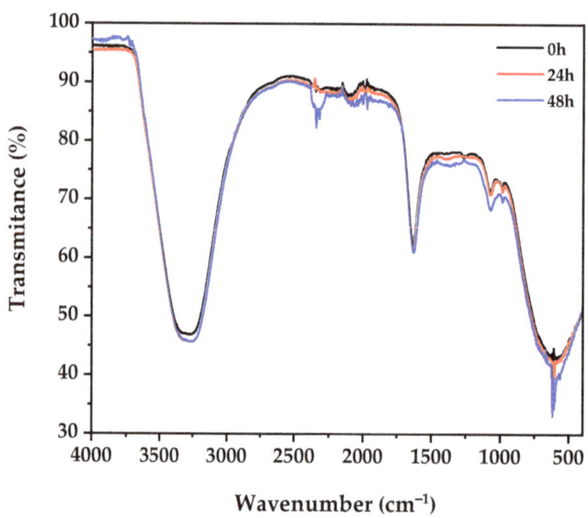

Figure 5. FTIR spectra of supernatant samples taken at initial time, 24, and 48 h.

Specifically, the peaks at 2120–2260 cm^{-1} corresponding to C≡N stretching vibrations [70] decreased over time. This directly reflects cyanide levels being metabolized by the bacteria. Concurrently, a broad peak around 3300 cm^{-1} is associated with OH and NH vibrations and this increased as the end product ammonia built up [71]. A third notable change is the growth of the peak at 1550–1750 cm^{-1} related to C=O vibrations from generated formate [72].

Taken together, the kinetic disappearance of cyanide bands and the emergence of ammonia and formate features provide strong spectroscopic evidence corroborating the proposed cyanide dihydratase mechanism. The second-order rate behavior reflects how this inducible enzymatic process controls the biodegradation pace.

4. Discussion

Cyanide is a highly toxic compound to living organisms that is used in various economic activities, mainly related to mining [73,74]. These mineral extraction activities generate wastes contaminated with heavy metals or toxic substances like CN that disperse into the environment [13,75]. This study isolated eight bacterial strains from a soil sample with sulfur mining residues that could tolerate 50–1000 ppm CN under alkaline conditions (Figure 1). Strain TT10s was identified as *Bacillus subtilis* and chosen for further study due to its exceptional ability to withstand and degrade high cyanide concentrations under alkaline conditions compared to prior *B. subtilis* strains and its rapid growth rate.

The identification of this species within a very diverse genus was accomplished through its phenotypic characteristics in relation to nutritional requirements, growth conditions (Tables 1 and 2), and DNA composition [76]. The 16S rRNA gene sequence confirms TT10s as a *B. subtilis* strain, sharing 100% nucleotide similarity. However, TT10s possesses unique functional capabilities, enabling rapid and thorough cyanide mineralization under alkaline conditions relevant for mining effluents.

Previous studies reported a 92–500 ppm cyanide tolerance of *B. subtilis* isolates [28,77], and TT10s survived up to 1000 ppm cyanide at pH 10.5. Additionally, TT10s achieved a complete degradation of 1000 ppm cyanide in just 48 h (Figure 3), far exceeding the 60–72 h timeframes for 500 ppm removal reported for other strains [28]. Other Bacillus species degraded lower CN concentrations, like *B. pumilus* up to 500 ppm [73]. Many bacterial genera like Pseudomonas, Rhodococcus, Klebsiella, and Bacillus can biodegrade CN, with specific strains listed in Table 4. These bacteria likely use metabolic pathways to completely degrade or transform CN into less harmful products [78,79], utilizing CN as nitrogen and carbon sources [34,80].

Degradation kinetics revealed a second order rate (k_2 = 0.08649 mg/(mg·h), R^2 = 0.96622) (Figure 4). The second-order rate law integrates both reactant (cyanide) and biocatalyst (bacterial cells) concentrations, consistent with the reaction being mediated by inducible bacterial enzymes that increase in amount with microbial growth [81]. Specifically, the cyanide dihydratase enzyme was synthesized by Bacillus subtilis in response to cyanide, which served as both inducer and substrate. Thus, the reaction rate depended on the concentrations of both cyanide and induced cyanide dihydratase, aligning with the second-order model [28].

Based on FTIR analysis (Figure 5), the hydrolytic pathway in *B. subtilis* produced ammonia and formate, characteristic of enzymatic CN triple bond cleavage mediated by water molecules [34,82]. *B. subtilis* convered this pollutant into less toxic substances [83]. Similar traits occurred in *B. safensis*, *B. cereus*, and *B. pumilus*, degrading CN enzymatically via cyanide dihydratase into ammonia and formate, with active enzymes up to pH 9 [22,84–86]. Some Bacillus species do not require cofactors for CN degradation [85,86]. The ability of *B. subtilis* to degrade CN without extra carbon sources at pH up to 10.5 shows its degradation enzyme functions under high pH levels, CN concentration, and without other nutrients.

Unlike other bacteria, that require additional nutrients to be more efficient in the biodegradation process [87,88], this bacterium performed the biodegradation process at a pH adjusted to 10.5 so that the CN remained dissolved in the medium rather than transforming into HCN [89]. This parameter provides an advantage and is one of the most important factors during biodegradation treatment [30,90]. Because Bacillus species are key producers of extracellular proteases, with potential applications to function and maintain stability under extreme alkaline conditions between pH 9–12 through sophisticated intracellular proton transport mechanisms for their growth [80], these proteases are induced for production by medium conditions like pH, temperature, and aeration, and strongly influenced by the available carbon and nitrogen components [81].

Using bacteria like *B. subtilis* for biodegradation with advantageous traits can enable applications in contaminated environments in an eco-friendly manner with minimal toxic byproducts [83]. Optimal conditions like temperature, pH, aeration, and nutrient availability should be ensured for efficient CN removal [74,91]. Bacterial biodegradation, with 85–98% efficiency, is more economical and eco-friendly than physical/chemical methods, generating minimal waste [73]. The highly efficient cyanide degradation capacity of the *Bacillus subtilis* strain under alkaline conditions holds significance for mitigating cyanide pollution from diverse industries like mining, textiles, metal finishing, and other industries that generate alkaline effluents [26]. Due to this, conventional remediation methods are hampered by variations in conditions, high costs, and hazardous byproducts [3].

At present, research efforts are crucial to fine-tune the biodegradation process of *B. subtilis* and elucidate its genetic underpinnings concerning cyanide degradation. The harnessing of *B. subtilis* holds immense potential for effective cyanide remediation in both aquatic and terrestrial ecosystems affected by contamination. This versatile bacterium presents an opportunity for the low-cost, environmentally friendly cleanup of cyanide-contaminated waters and soils, particularly at lower pH ranges. The progress made in this research sets a pivotal foundation for the development of a sustainable and eco-friendly biotechnological solution, addressing the pressing challenges posed by cyanide pollution.

5. Conclusions

This study isolated a novel *Bacillus subtilis* strain designated TT10s from mining environmental liabilities that demonstrated efficient cyanide biodegradation capacity.

The analysis of the 16S rRNA gene sequence established the phylogenetic classification of TT10s as *Bacillus subtilis*. This novel isolate displayed rapid adaptation and cyanide degradation under alkaline conditions with no need for extra carbon source conditions.

The biodegradation experiments showed that *Bacillus subtilis* demonstrated exceptional cyanide degradation capacity under alkaline conditions of pH 10.5, achieving the 100% removal of an initial cyanide concentration of 1000 ppm in less than 48 h.

Kinetic experiments revealed that the cyanide degradation followed second-order rate kinetics (k_2 = 0.08649 mg/(mg·h) and R^2 = 0.96622), with dependence on both cyanide concentration and bacterial density.

The quantitative kinetics aligned with an enzymatic mechanism governed by cyanide dihydratase, which converts cyanide into ammonia and formate. FTIR analysis provided additional spectroscopic evidence supporting this enzymatic reaction. The kinetic and FTIR data implicate cyanide dihydratase as the rate-controlling enzyme in alkaline cyanide biodegradation by TT10s.

Author Contributions: Conceptualization C.J.C.Q. and G.d.L.F.Q.; methodology, G.d.L.F.Q.; software, G.d.L.F.Q. and E.J.S.S.; validation, C.J.C.Q., G.d.L.F.Q. and E.J.S.S.; formal analysis, C.J.C.Q.; investigation, C.J.C.Q. and G.d.L.F.Q.; resources, C.J.C.Q.; data curation, G.d.L.F.Q.; writing—original draft preparation, C.J.C.Q., G.d.L.F.Q. and E.J.S.S.; writing—review and editing, C.J.C.Q., G.d.L.F.Q., E.J.S.S., M.C.M. and G.J.M.C.; visualization, G.d.L.F.Q.; supervision, C.J.C.Q.; project administration, G.d.L.F.Q.; funding acquisition, C.J.C.Q. All authors have read and agreed to the published version of the manuscript.

Funding: This research and APC were funded by the Universidad Nacional Jorge Basadre Grohmann through "Fondos del canon sobrecanon y regalias mineras", approved by Rectoral Resolution N° 4723-2015-UN/JBG, with the project "Análisis genómico de microorganismos degradadores de cianuro para la remediación, en fase de biorreactor de pasivos ambientales mineros de la región de Tacna".

Data Availability Statement: The data presented in this study are available within this article.

Acknowledgments: We are grateful to the project "Determination of the optical fingerprints of solid, liquid and organic materials using visible and infrared spectroscopy", of the Universidad Nacional

Jorge Basadre Grohmann, approved by rectoral resolutions N° 5854-2019-UN/JBG, for their support in the FTIR analysis.

Conflicts of Interest: The authors declare no conflict of interest.

References

1. Wang, Z.-S.; Cho, Y.-C.; Lin, Y.-P. Removal of Cyanide from Electroplating Wastewater by Persulfate Oxidation Process. 2023. Available online: https://ssrn.com/abstract=4427783 (accessed on 28 September 2023).
2. Eisler, R.; Wiemeyer, S.N. Cyanide hazards to plants and animals from gold mining and related water issues. *Rev. Environ. Contam. Toxicol.* **2004**, *183*, 21–54. [CrossRef] [PubMed]
3. Martínková, L.; Bojarová, P.; Sedova, A.; Křen, V. Recent trends in the treatment of cyanide-containing effluents: Comparison of different approaches. *Crit. Rev. Environ. Sci. Technol.* **2023**, *53*, 416–434. [CrossRef]
4. Cacciuttolo, C.; Cano, D. Environmental Impact Assessment of Mine Tailings Spill Considering Metallurgical Processes of Gold and Copper Mining: Case Studies in the Andean Countries of Chile and Peru. *Water* **2022**, *14*, 3057. [CrossRef]
5. Matlock, M.M.; Howerton, B.S.; van Aelstyn, M.A.; Nordstrom, F.L.; Atwood, D.A. Advanced mercury removal from gold leachate solutions prior to gold and silver extraction: A field study from an active gold mine in Peru. *Environ. Sci. Technol.* **2002**, *36*, 1636–1639. [CrossRef]
6. Isla, A. The Guardians of Conga Lagoons: Defending Land Water and Freedom in Peru. *Can. Woman Stud./Les Cah. De La Femme* **2015**, *30*, 25–40.
7. Presidencia del Concejo de Ministros de Perú. Decreto Supremo que Declara el Estado de Emergencia en los Distritos de Cocachacra, Dean Valdivia y Punta de Bombón de la Provincia de Islay, del Departamento de Arequipa, por Peligro Inminente Ante Contaminación Hídrica; Perú, 2021 (N° 106-2021-PCM). Available online: https://busquedas.elperuano.pe/dispositivo/NL/1957547-2 (accessed on 26 September 2023).
8. Bebbington, A.; Williams, M. Water and Mining Conflicts in Peru. *Mt. Res. Dev.* **2008**, *28*, 190–195. [CrossRef]
9. Salem, J.; Amonkar, Y.; Maennling, N.; Lall, U.; Bonnafous, L.; Thakkar, K. An analysis of Peru: Is water driving mining conflicts? *Resour. Policy* **2021**, *74*, 101270. [CrossRef]
10. Ministerio del Ambiente. Modifican los Estándares Nacionales de Calidad Ambiental Para Agua y Establecen Disposiciones Complementarias Para su Aplicación; Diario el Peruano, 2015 (N° 015-2015-MINAM). Available online: https://www.minam.gob.pe/wp-content/uploads/2015/12/Decreto-Supremo-N%C2%B0-015-2015-MINAM.pdf (accessed on 28 September 2023).
11. Marshall, B.G.; Veiga, M.M.; Da Silva, H.A.M.; Guimarães, J.R.D. Cyanide Contamination of the Puyango-Tumbes River Caused by Artisanal Gold Mining in Portovelo-Zaruma, Ecuador. *Curr. Environ. Health Rep.* **2020**, *7*, 303–310. [CrossRef]
12. Copari Mamani, A.B.; Carpio Mamani, M.; Cáceda Quiroz, C.J. Optimización de factores fisicoquímicos en la biodegradación de cianuro por Klebsiella sp.ART1, en biorreactor aireado. *Cienc. Desarro.* **2020**, *26*, 20–31. [CrossRef]
13. Quiroz, C.J.C.; Choque, G.J.M.; Mamani, M.C.; Quispe, G.D.L.F. Evaluation of the content of metals and contamination indices generated by environmental liabilities, in Tacna, Peru. *Res. Sq.* **2022**. [CrossRef]
14. Swenson, J.J.; Carter, C.E.; Domec, J.-C.; Delgado, C.I. Gold mining in the Peruvian Amazon: Global prices, deforestation, and mercury imports. *PLoS ONE* **2011**, *6*, e18875. [CrossRef]
15. Veiga, M.M.; Angeloci, G.; Hitch, M.; Colon Velasquez-Lopez, P. Processing centres in artisanal gold mining. *J. Clean. Prod.* **2014**, *64*, 535–544. [CrossRef]
16. Mayorca Clemente, S. *Reducción de Cianuro Del Agua Industrial Contaminada Mediante Biopelícula Microbiana, Chala—Arequipa 2018*; Universidad César Vallejo: Lima, Peru, 2018.
17. Vuono, D.C.; Vanneste, J.; Figueroa, L.A.; Hammer, V.; Aguilar-Huaylla, F.N.; Malone, A.; Smith, N.M.; Garcia-Chevesich, P.A.; Bolaños-Sosa, H.G.; Alejo-Zapata, F.D.; et al. Photocatalytic Advanced Oxidation Processes for Neutralizing Free Cyanide in Gold Processing Effluents in Arequipa, Southern Peru. *Sustainability* **2021**, *13*, 9873. [CrossRef]
18. Asner, G.P.; Tupayachi, R. Accelerated losses of protected forests from gold mining in the Peruvian Amazon. *Environ. Res. Lett.* **2016**, *12*, 94004. [CrossRef]
19. Downey, J.; Basi, K.; DeFreytas, M.; Rockwood, G.; Chronic cyanide exposure. *Case Studies and Animal Models in Toxicology of Cyanides and Cyanogens: Experimental, Applied and Clinical Aspects*, 1st ed.; Hall, A.H., Isom, G.E., Rockwood, G.A., Eds.; John Wiley & Sons: Chichester, UK, 2015; pp. 21–40. ISBN 978-1119978534.
20. Shifrin, N.S.; Beck, B.D.; Gauthier, T.D.; Chapnick, S.D.; Goodman, G. Chemistry, toxicology, and human health risk of cyanide compounds in soils at former manufactured gas plant sites. *Regul. Toxicol. Pharmacol.* **1996**, *23*, 106–116. [CrossRef]
21. Kuyucak, N.; Akcil, A. Cyanide and removal options from effluents in gold mining and metallurgical processes. *Miner. Eng.* **2013**, *50–51*, 13–29. [CrossRef]
22. Sharma, M.; Akhter, Y.; Chatterjee, S. A review on remediation of cyanide containing industrial wastes using biological systems with special reference to enzymatic degradation. *World J. Microbiol. Biotechnol.* **2019**, *35*, 70. [CrossRef]
23. Akcil, A. Destruction of cyanide in gold mill effluents: Biological versus chemical treatments. *Biotechnol. Adv.* **2003**, *21*, 501–511. [CrossRef]
24. Luque-Almagro, V.M.; Moreno-Vivián, C.; Roldán, M.D. Biodegradation of cyanide wastes from mining and jewellery industries. *Curr. Opin. Biotechnol.* **2016**, *38*, 9–13. [CrossRef]

25. Kunz, D.A.; Wang, C.S.; Chen, J.L. Alternative routes of enzymic cyanide metabolism in Pseudomonas fluorescens NCIMB 11764. *Microbiology* **1994**, *140 Pt 7*, 1705–1712. [CrossRef]
26. Dash, R.R.; Gaur, A.; Balomajumder, C. Cyanide in industrial wastewaters and its removal: A review on biotreatment. *J. Hazard. Mater.* **2009**, *163*, 1–11. [CrossRef] [PubMed]
27. Castric, P.A.; Strobel, G.A. Cyanide Metabolism by Bacillus megaterium. *J. Biol. Chem.* **1969**, *244*, 4089–4094. [CrossRef]
28. Rosario, C.G.A.; Vallenas-Arévalo, A.T.; Arévalo, S.J.; Espinosa, D.C.R.; Tenório, J.A.S. Biodegradation of cyanide using a Bacillus subtilis strain isolated from artisanal gold mining tailings. *Braz. J. Chem. Eng.* **2023**, *40*, 129–136. [CrossRef]
29. Meyers, P.R.; Gokool, P.; Rawlings, D.E.; Woods, D.R. An efficient cyanide-degrading Bacillus pumilus strain. *J. Gen. Microbiol.* **1991**, *137*, 1397–1400. [CrossRef] [PubMed]
30. Alvillo-Rivera, A.; Garrido-Hoyos, S.; Buitrón, G.; Thangarasu-Sarasvathi, P.; Rosano-Ortega, G. Biological treatment for the degradation of cyanide: A review. *J. Mater. Res. Technol.* **2021**, *12*, 1418–1433. [CrossRef]
31. Liu, J.K.; Liu, C.H.; Lin, C.S. The role of nitrogenase in a cyanide-degrading Klebsiella oxytoca strain. *Proc. Natl. Sci. Counc. Repub. China B* **1997**, *21*, 37–42. [PubMed]
32. Adjei, M.D.; Ohta, Y. Isolation and characterization of a cyanide-utilizing Burkholderia cepacia strain. *World J. Microbiol. Biotechnol.* **1999**, *15*, 699–704. [CrossRef]
33. Nallapan Maniyam, M.; Sjahrir, F.; Ibrahim, A.L.; Cass, A.E.G. Biodegradation of cyanide by acetonitrile-induced cells of Rhodococcus sp. UKMP-5M. *J. Gen. Appl. Microbiol.* **2013**, *59*, 393–404. [CrossRef]
34. Ebbs, S. Biological degradation of cyanide compounds. *Curr. Opin. Biotechnol.* **2004**, *15*, 231–236. [CrossRef]
35. Nwokoro, O.; Dibua, M.E.U. Degradation of soil cyanide by single and mixed cultures of Pseudomonas stutzeri and Bacillus subtilis. *Arh. Hig. Rada Toksikol.* **2014**, *65*, 113–119. [CrossRef]
36. Niu, H.; Volesky, B. Characteristics of gold biosorption from cyanide solution. *J. Chem. Technol. Biotechnol.* **1999**, *74*, 778–784. [CrossRef]
37. Knowles, C.J. Cyanide utilization and degradation by microorganisms. *Ciba Found. Symp.* **1988**, *140*, 3–15. [CrossRef] [PubMed]
38. Patil, Y.; Paknikar, K. Development of a process for biodetoxification of metal cyanides from waste waters. *Process Biochem.* **2000**, *35*, 1139–1151. [CrossRef]
39. Obed Ntwampe, S.K. Biodegradation of Free Cyanide Using Bacillus Sp. Consortium Dominated by Bacillus Safensis, Licheniformis and Tequilensis Strains: A Bioprocess Supported Solely with Whey. *J. Bioremediation Biodegrad.* **2014**, *5*. [CrossRef]
40. Akcil, A.; Karahan, A.; Ciftci, H.; Sagdic, O. Biological treatment of cyanide by natural isolated bacteria (*Pseudomonas* sp.). *Miner. Eng.* **2003**, *16*, 643–649. [CrossRef]
41. Nallapan Maniyam, M.; Sjahrir, F.; Ibrahim, A.L.; Cass, A.E.G. Cyanide degradation by immobilized cells of Rhodococcus UKMP-5M. *Biologia* **2012**, *67*, 837–844. [CrossRef]
42. Ozkan, M.; Desai, S.G.; Zhang, Y.; Stevenson, D.M.; Beane, J.; White, E.A.; Guerinot, M.L.; Lynd, L.R. Characterization of 13 newly isolated strains of anaerobic, cellulolytic, thermophilic bacteria. *J. Ind. Microbiol. Biotechnol.* **2001**, *27*, 275–280. [CrossRef]
43. Mirizadeh, S.; Yaghmaei, S.; Ghobadi Nejad, Z. Biodegradation of cyanide by a new isolated strain under alkaline conditions and optimization by response surface methodology (RSM). *J. Environ. Health Sci. Eng.* **2014**, *12*, 85. [CrossRef]
44. Kumar, V.; Kumar, V.; Bhalla, T.C. Statistical Enhancement of Cyanide Degradation Using Microbial Consortium. *J. Microb. Biochem. Technol.* **2015**, *7*, 6. [CrossRef]
45. Huertas, M.J.; Sáez, L.P.; Roldán, M.D.; Luque-Almagro, V.M.; Martínez-Luque, M.; Blasco, R.; Castillo, F.; Moreno-Vivián, C.; García-García, I. Alkaline cyanide degradation by Pseudomonas pseudoalcaligenes CECT5344 in a batch reactor. Influence of pH. *J. Hazard. Mater.* **2010**, *179*, 72–78. [CrossRef]
46. Rice, E.W.; Baird, R.B.; Eaton, A.D. *Standard Methods For the Examination of Water and Wastewater*; American Public Health Association: Washington, DC, USA, 2017; ISBN 978-0-87553-299-8.
47. Kandasamy, S.; Dananjeyan, B.; Krishnamurthy, K.; Benckiser, G. Aerobic cyanide degradation by bacterial isolates from cassava factory wastewater. *Braz. J. Microbiol.* **2015**, *46*, 659–666. [CrossRef] [PubMed]
48. Alvarado-López, M.J.; Garrido-Hoyos, S.E.; Raynal-Gutiérrez, M.E.; El-Kassis, E.G.; Luque-Almagro, V.M.; Rosano-Ortega, G. Cyanide Biodegradation by a Native Bacterial Consortium and Its Potential for Goldmine Tailing Biotreatment. *Water* **2023**, *15*, 1595. [CrossRef]
49. Ibrahim, K.K.; Syed, M.A.; Shukor, M.Y.; Ahmad, S.A. Biological Remediation of Cyanide: A Review. *Biotropia* **2016**, *22*, 151–163. [CrossRef]
50. Mekuto, L.; Alegbeleye, O.O.; Ntwampe, S.K.O.; Ngongang, M.M.; Mudumbi, J.B.; Akinpelu, E.A. Co-metabolism of thiocyanate and free cyanide by Exiguobacterium acetylicum and Bacillus marisflavi under alkaline conditions. *3 Biotech* **2016**, *6*, 173. [CrossRef]
51. Mekuto, L.; Ntwampe, S.K.O.; Jackson, V.A. Biodegradation of free cyanide and subsequent utilisation of biodegradation by-products by Bacillus consortia: Optimisation using response surface methodology. *Environ. Sci. Pollut. Res. Int.* **2015**, *22*, 10434–10443. [CrossRef]
52. Wu, C.-F.; Xu, X.-M.; Zhu, Q.; Deng, M.-C.; Feng, L.; Peng, J.; Yuan, J.-P.; Wang, J.-H. An effective method for the detoxification of cyanide-rich wastewater by Bacillus sp. CN-22. *Appl. Microbiol. Biotechnol.* **2014**, *98*, 3801–3807. [CrossRef]
53. Dubey, S.K.; Holmes, D.S. Biological cyanide destruction mediated by microorganisms. *World J. Microbiol. Biotechnol.* **1995**, *11*, 257–265. [CrossRef]

54. Slepecky, R.A.; Hemphill, H.E. The Genus Bacillus—Nonmedical. In *The Prokaryotes*; Dworkin, M., Falkow, S., Rosenberg, E., Schleifer, K.-H., Stackebrandt, E., Eds.; Springer: New York, NY, USA, 2006; pp. 530–562. ISBN 978-0-387-25494-4.
55. Stülke, J.; Hillen, W. Regulation of carbon catabolism in Bacillus species. *Annu. Rev. Microbiol.* **2000**, *54*, 849–880. [CrossRef]
56. Berkeley, R.; Logan, N.A.; Shute, L.A.; Capey, A.G. *12 Identification of Bacillus Species*; Elsevier: Amsterdam, The Netherlands, 1984; pp. 291–328. ISBN 9780125215169.
57. Bhattacharya, D.; de Los Santos Villalobos, S.; Ruiz, V.V.; Selvin, J.; Mukherjee, J. *Bacillus rugosus* sp. nov. producer of a diketopiperazine antimicrobial, isolated from marine sponge *Spongia officinalis* L. *Antonie Van Leeuwenhoek* **2020**, *113*, 1675–1687. [CrossRef]
58. Gatson, J.W.; Benz, B.F.; Chandrasekaran, C.; Satomi, M.; Venkateswaran, K.; Hart, M.E. *Bacillus tequilensis* sp. nov., isolated from a 2000-year-old Mexican shaft-tomb, is closely related to Bacillus subtilis. *Int. J. Syst. Evol. Microbiol.* **2006**, *56*, 1475–1484. [CrossRef]
59. Dunlap, C.A.; Bowman, M.J.; Zeigler, D.R. Promotion of Bacillus subtilis subsp. inaquosorum, Bacillus subtilis subsp. spizizenii and Bacillus subtilis subsp. stercoris to species status. *Antonie Van Leeuwenhoek* **2020**, *113*, 1–12. [CrossRef] [PubMed]
60. Muloiwa, M.; Nyende-Byakika, S.; Dinka, M. Comparison of unstructured kinetic bacterial growth models. *S. Afr. J. Chem. Eng.* **2020**, *33*, 141–150. [CrossRef]
61. Malone, S.L.; Pearce, L.L.; Peterson, J. Environmental toxicology of cyanide. In *Toxicology of Cyanides and Cyanogens: Experimental, Applied and Clinical Aspects*, 1st ed.; Hall, A.H., Isom, G.E., Rockwood, G.A., Eds.; John Wiley & Sons: Chichester, UK, 2015; pp. 82–97. ISBN 978-1119978534.
62. Kunz, D.A.; Nagappan, O.; Silva-Avalos, J.; Delong, G.T. Utilization of cyanide as nitrogenous substrate by Pseudomonas fluorescens NCIMB 11764: Evidence for multiple pathways of metabolic conversion. *Appl. Environ. Microbiol.* **1992**, *58*, 2022–2029. [CrossRef] [PubMed]
63. Avcioglu, N.H.; Bilkay, I.S. Biological Treatment of Cyanide by Using Klebsiella pneumoniae Species. *Food Technol. Biotechnol.* **2016**, *54*, 450–454. [CrossRef] [PubMed]
64. Dwivedi, N.; Balomajumder, C.; Mondal, P. Comparative evaluation of cyanide removal by adsorption, biodegradation, and simultaneous adsorption and biodegradation (SAB) process using Bacillus cereus and almond shell. *J. Environ. Biol.* **2016**, *37*, 551.
65. Justo Arevalo, S.; Zapata Sifuentes, D.; Cuba Portocarrero, A.; Brescia Reátegui, M.; Monge Pimentel, C.; Farage Martins, L.; Marques Pierry, P.; Morais Piroupo, C.; Guerra Santa Cruz, A.; Quiñones Aguilar, M.; et al. Genomic Characterization of Bacillus safensis Isolated from Mine Tailings in Peru and Evaluation of Its Cyanide-Degrading Enzyme CynD. *Appl. Environ. Microbiol.* **2022**, *88*, e0091622. [CrossRef]
66. Watanabe, A.; Yano, K.; Ikebukuro, K.; Karube, I. Cyanide hydrolysis in a cyanide-degrading bacterium, Pseudomonas stutzeri AK61, by cyanidase. *Microbiology* **1998**, *144 Pt 6*, 1677–1682. [CrossRef]
67. Panay, A.J.; Vargas-Serna, C.L.; Carmona-Orozco, M.L. Biodegradation of cyanide using recombinant Escherichia coli expressing Bacillus pumilus cyanide dihydratase. *Rev. Colomb. Biotecnol.* **2020**, *22*, 27–35. [CrossRef]
68. Dursun, A.Y.; Uslu, G.; Cuci, Y.; Aksu, Z. Bioaccumulation of copper(II), lead(II) and chromium(VI) by growing Aspergillus niger. *Process Biochem.* **2003**, *38*, 1647–1651. [CrossRef]
69. Maciel, A.C.; Da Silva Pena, R.; do Nascimento, L.D.; de Oliveira, T.A.; Chagas-Junior, G.C.A.; Lopes, A.S. Health exposure risks and bioremediation of cyanide in cassava processing effluents: An overview. *J. Water Process Eng.* **2023**, *55*, 104079. [CrossRef]
70. Monga, I.; Paul, V.; Muniyasamy, S.; Zinyemba, O. Green Synthesis of Sodium Cyanide Using Hydrogen Cyanide Extracted under Vacuum from Cassava (Manihot esculenta Crantz) Leaves. *Sustain. Chem.* **2022**, *3*, 312–333. [CrossRef]
71. Shivanoor, S.M.; David, M. Fourier transform infrared (FT-IR) study on cyanide induced biochemical and structural changes in rat sperm. *Toxicol. Rep.* **2015**, *2*, 1347–1356. [CrossRef]
72. Nandiyanto, A.B.D.; Oktiani, R.; Ragadhita, R. How to Read and Interpret FTIR Spectroscope of Organic Material. *Indones. J. Sci. Technol.* **2019**, *4*, 97. [CrossRef]
73. Ojaghi, A.; Shafaie Tonkaboni, S.Z.; Shariati, P.; Doulati Ardejani, F. Novel cyanide electro-biodegradation using Bacillus pumilus ATCC 7061 in aqueous solution. *J. Environ. Health Sci. Eng.* **2018**, *16*, 99–108. [CrossRef] [PubMed]
74. Razanamahandry, L.C.; Andrianisa, H.A.; Karoui, H.; Kouakou, K.M.; Yacouba, H. Biodegradation of free cyanide by bacterial species isolated from cyanide-contaminated artisanal gold mining catchment area in Burkina Faso. *Chemosphere* **2016**, *157*, 71–78. [CrossRef]
75. Abouian Jahromi, M.; Jamshidi-Zanjani, A.; Khodadadi Darban, A. Heavy metal pollution and human health risk assessment for exposure to surface soil of mining area: A comprehensive study. *Environ. Earth Sci.* **2020**, *79*, 365. [CrossRef]
76. Celandroni, F.; Vecchione, A.; Cara, A.; Mazzantini, D.; Lupetti, A.; Ghelardi, E. Identification of Bacillus species: Implication on the quality of probiotic formulations. *PLoS ONE* **2019**, *14*, e0217021. [CrossRef]
77. Durairaju Nisshanthini, S.; Teresa Infanta S, A.K.; Raja, D.S.; Natarajan, K.; Palaniswamy, M.; Angayarkanni, J. Spectral characterization of a pteridine derivative from cyanide-utilizing bacterium Bacillus subtilis—JN989651. *J. Microbiol.* **2015**, *53*, 262–271. [CrossRef]
78. Dangi, A.K.; Sharma, B.; Hill, R.T.; Shukla, P. Bioremediation through microbes: Systems biology and metabolic engineering approach. *Crit. Rev. Biotechnol.* **2019**, *39*, 79–98. [CrossRef]

79. Luque-Almagro, V.M.; Huertas, M.-J.; Martínez-Luque, M.; Moreno-Vivián, C.; Roldán, M.D.; García-Gil, L.J.; Castillo, F.; Blasco, R. Bacterial degradation of cyanide and its metal complexes under alkaline conditions. *Appl. Environ. Microbiol.* **2005**, *71*, 940–947. [CrossRef]
80. Satyanarayana, T.; Johri, B.N. *Microorganisms in Environmental Management*; Springer: Dordrecht, The Netherlands, 2012; ISBN 978-94-007-2228-6.
81. Kaspar, F.; Neubauer, P.; Gimpel, M. Bioactive Secondary Metabolites from Bacillus subtilis: A Comprehensive Review. *J. Nat. Prod.* **2019**, *82*, 2038–2053. [CrossRef] [PubMed]
82. Luque Almagro, V.M. *Metabolismo Del Cianuro y Del Cianato en "Pseudomonas pseudoalcaligenes" CECT5344: Aplicaciones Biotecnológicas: (Accésit)*; Analistas Económicos de Andalucía: Málaga, Spain, 2006; ISBN 978-84-95191-81-6.
83. Botz, M.M.; Mudder, T.I.; Akcil, A.U. Cyanide Treatment. In *Gold Ore Processing*; Elsevier: Amsterdam, The Netherlands, 2016; pp. 619–645. ISBN 9780444636584.
84. Benedik, M.J.; Sewell, B.T. Cyanide-degrading nitrilases in nature. *J. Gen. Appl. Microbiol.* **2018**, *64*, 90–93. [CrossRef] [PubMed]
85. Crum, M.A.; Sewell, B.T.; Benedik, M.J. Bacillus pumilus Cyanide Dihydratase Mutants with Higher Catalytic Activity. *Front. Microbiol.* **2016**, *7*, 1264. [CrossRef] [PubMed]
86. Jandhyala, D.; Berman, M.; Meyers, P.R.; Sewell, B.T.; Willson, R.C.; Benedik, M.J. CynD, the cyanide dihydratase from Bacillus pumilus: Gene cloning and structural studies. *Appl. Environ. Microbiol.* **2003**, *69*, 4794–4805. [CrossRef]
87. Guamán Guadalima, M.P.; Nieto Monteros, D.A. Evaluation of the rotational speed and carbon source on the biological removal of free cyanide present on gold mine wastewater, using a rotating biological contactor. *J. Water Process Eng.* **2018**, *23*, 84–90. [CrossRef]
88. Terada, A.; Komatsu, D.; Ogawa, T.; Flamandita, D.; Sahlan, M.; Nishimura, M.; Yohda, M. Isolation of cyanide-degrading bacteria and molecular characterization of its cyanide-degrading nitrilase. Biotechnol. *Appl. Biochem.* **2022**, *69*, 183–189. [CrossRef]
89. Oudjehani, K.; Zagury, G.J.; Deschênes, L. Natural attenuation potential of cyanide via microbial activity in mine tailings. *Appl. Microbiol. Biotechnol.* **2002**, *58*, 409–415. [CrossRef]
90. Deloya, A. Tratamiento de desechos del cianuro por biorremediación. *Tecnología en Marcha* **2012**, *25*, 61–72. [CrossRef]
91. Baxter, J.; Cummings, S.P. The current and future applications of microorganism in the bioremediation of cyanide contamination. *Antonie Van Leeuwenhoek* **2006**, *90*, 1–17. [CrossRef]

water

MDPI

Article

Redox Behavior of Chromium in the Reduction, Coagulation, and Biotic Filtration (RCbF) Drinking Water Treatment—A Pilot Study

Daniel Mahringer [1,*], Sami S. Zerelli [1,2] and Aki S. Ruhl [1,3]

1 German Environment Agency (UBA), Section II 3.3, Schichauweg 58, 12307 Berlin, Germany; akisebastian.ruhl@uba.de (A.S.R.)
2 Gesellschaft zur Förderung der Naturwissenschaftlich-Technischen Forschung, Volmerstr. 7B, 12489 Berlin, Germany
3 Water Treatment, Faculty III—Process Sciences, Technische Universität Berlin, Sekr. KF4, Straße des 17. Juni 135, 10623 Berlin, Germany
* Correspondence: daniel.mahringer@uba.de

Abstract: The chromium (Cr) limit values are currently tightened to 25 μg L^{-1} (EU), 5 μg L^{-1} (Germany), and possibly 10 μg L^{-1} Cr(VI) (California). The combined process of chemical reduction, coagulation, and biotic filtration (RCbF) efficiently removes Cr(VI) in drinking water. In this study, redox-active substances (O_2, NO_3^-, Fe^{2+}, MnO_2) were investigated concerning their effect on the RCbF process. The experiments were performed at two-stage pilot waterworks for biological iron and manganese removal. O_2 or NO_3^- as oxidants affected the RCbF process, neither by consumption of the reductant Fe(II) nor by re-oxidation of already formed Cr(III) in the supernatant of the filter bed. However, the oxidation of Cr(III) by O_2 to Cr(VI) with MnO_2 as a mediator was identified as potential risk for Cr breakthrough. Up to one third of the initial Cr(III) concentration was oxidized to Cr(VI) in the second filter bed within a contact time of only 5 min. The kinetically relevant mechanism seemed to be the formation of Cr(III)Fe(III)-hydroxides and not the reduction of Cr(VI) by Fe(II). Further, the mixing of Cr(VI) containing raw water with Fe(II) containing groundwater was determined as a chemical-free alternative for the RCbF process, depending on the resulting Fe(II) concentration after mixing.

Keywords: hexavalent chromium; chromate removal; trivalent chromium; drinking water treatment; reduction coagulation biological filtration

1. Introduction

Chromium (Cr) is a ubiquitously occurring heavy metal in the environment [1]. Its natural appearance in water is limited to the two oxidation states hexavalent chromium (Cr(VI)) and trivalent chromium (Cr(III)) as amphoteric species of chromium hydroxide with much lower solubility and mobility than Cr(VI) [2–5].

There is an ongoing debate among experts about the toxicity of Cr(VI) in drinking water and its mode of action within the human body. While by some a mutagenic mode of action is favored [6–9], others assume a threshold mode of action [10–15] for Cr(VI)-induced carcinogenesis. The latter approach is favored by the World Health Organization (WHO), which assessed data from animal studies by the National Toxicology Program [16] and concluded that hyperplasia in the small intestine for tumor development is to be taken as the most sensitive end point [17]. If a mutagenic mode of action is applied, an assessment in which the data from animal experiments are extrapolated to a lifetime exposure risk for humans would result in a much lower limit value in drinking water [18].

In the latest EU Drinking Water Directive, the limitation value for total chromium (Cr_{tot}) was lowered from 50 μg L^{-1} to 25 μg L^{-1}, whereas the enforcement of this value is postponed to 2036 [19]. In Germany, a limitation value of 25 $\mu g\ L^{-1}$ is directly implemented, and further tightening to a limitation value of 5 $\mu g\ L^{-1}$ is scheduled for 2030 [20]. In California, as a progressive state for Cr(VI) regulation, a limitation value of 10 $\mu g\ L^{-1}$ is proposed by the California Water Boards [21], and therefore this would tighten the current state maximum contaminant level of 50 $\mu g\ L^{-1}$ Cr_{tot} (the federal level of 100 $\mu g\ L^{-1}$ Cr_{tot}) currently defined in drinking water.

For Cr removal, ion exchange with weak basic anion exchange resins showed sufficient removal at a demonstration scale [22]. However, the water matrix, especially with sulfate as a competing anion, was identified as crucial for the performance [23]. Furthermore, a combination of reduction, coagulation, and filtration (RCF), with divalent iron (Fe(II)) as a reducing agent, showed good performance for Cr(VI) removal at the bench, pilot, and demonstration scales [22,24–26]. The RCF process was further developed by substituting the filtration with a biotic filtration (RCbF), which increased the effectiveness considerably [27]. In the RCbF process, dosed Fe(II) quickly reduces Cr(VI) in the supernatant of the filter bed to Cr(III), which is co-precipitated as an amorphous chromium(III)–iron(III) hydroxide solid solution with a low solubility [5]. This hydroxide is removed in the subsequent biological filter bed, where the excess Fe(II) is also removed very efficiently [27]. The Cr(VI) reduction and the formation of the hydroxide must be completed before entering the filter bed, because the reducing agent Fe(II) is (purposefully) oxidized to Fe(III) very quickly within the filter bed via a contact catalysis and biotic oxidation [28].

In this study, the oxidation/reduction potential (ORP)-dependent behavior of Cr in the RCbF process was investigated. Therefore, the reduction of Cr(VI) with Fe(II) was examined in the presence of competing oxidizing agents O_2 and NO_3^-, which potentially disturb the reduction of Cr(VI). In addition, the re-oxidation of Cr(III) in the process poses a risk for diminished efficiency of the RCbF process, either in the supernatant of the biological filter bed or even after the filter bed, if Cr(III) breaks through the filter. Therefore, the oxidation of Cr(III) with O_2 at a MnO_2 surface within a second filter bed was investigated with different O_2 doses. Further, since the RCbF process requires Fe(II) as a reducing agent, a chemical-free alternative in which Cr(VI)-containing raw water was mixed with Fe(II)-containing groundwater was investigated. The novelty of this study lies in the further investigation of the redox behavior of Cr in the RCbF process on the basis of process engineering variations at a semi-technical scale.

2. Materials and Methods

2.1. Chemicals

Sodium chromate, chromium trichloride hexahydrate, ferrous sulphate heptahydrate, sodium nitrate, potassium dihydrogenphosphate, hydrochloric acid (32%), sodium hydroxide (5 M), nitric acid (65%), and sodium chloride were purchased from Merck (Darmstadt, Germany). All chemicals were analytical grade, except nitric acid which was suprapur. Stock solutions were produced in relatively high concentrations and stored in a fridge. Dosing solution in big volumes were produced next to the pilot waterworks on a daily basis. Cr(VI) stock solution (1 $g\ L^{-1}$) was prepared by dissolving 3.20 g of sodium chromate in 1 L ultrapure water. Cr(VI) dosing solutions of 3.0 $mg\ L^{-1}$ were prepared by transferring 300 mL of the stock solution into 100 L of ultrapure water. Cr(III) stock solution (1 $g\ L^{-1}$) was prepared on a daily basis by dissolving 5.1 g of chromium trichloride hexahydrate in 1 L of ultrapure water. Cr(III) dosing solutions of 3.0 $mg\ L^{-1}$ were prepared by transferring 180 mL of stock solution into 60 L of ultrapure water. Nitrate dosing solution (12.5 $g\ L^{-1}$) was prepared by adding 1.25 kg of sodium nitrate into the Cr(VI) dosing solution. Moreover, 375 $mg\ L^{-1}$ Fe(II) dosing solutions were prepared by dissolving 187.5 g of ferrous sulphate heptahydrate in 100 L of ultrapure water after adjusting the pH to 3.0. For the stabilization of Cr(VI) samples, a phosphate buffer stock solution was prepared by dissolving 17.42 g of potassium dihydrogenphosphate in 200 mL of ultrapure water.

2.2. Pilot Waterworks

Fully automated two-stage pilot waterworks for biological iron and manganese removal (Figure S1, Supplementary Material) were used for the experiments. The waterworks are described in detail elsewhere [27,29].

The continuously operated waterworks usually treat reduced groundwater with 2.4 mg L^{-1} Fe(II) and 0.5 mg L^{-1} Mn(II). The groundwater is oxygenated by mass flow controllers (In-Flow controllers, Bronkhorst, Kamen, Germany) before each filter bed. In filter bed 1 Fe(II) and filter bed 2, Mn(II) is biotically oxidized, precipitated, and retained in the biofilm on the filter material. Before filter 1, O_2 dosage is controlled according to the target redox-potential of 30 mV before filter 2 in order to avoid precipitation of MnO_2 already in filter 1Before filter 2, O_2 dosage is manually adjusted, whereas the O_2 concentration after filter 2 should not exceed 2 mg L^{-1}.

The residence times in the waterworks were determined by a tracer test with NaCl and conductivity measurements, and a 50% breakthrough of conductivity was defined as the residence time until the respective sampling point.

2.3. Experimental Procedure

Before each experiment, both filter columns were backwashed, and raw water was switched from groundwater to oxygen-free drinking water from the technical-scale waterworks in Berlin-Marienfelde, Germany (constituents are shown in Table S1, Supplementary Material). Experiments took one filtration cycle (ca. 50 h) each, and samples were taken on the first and last day of the experiments from various sample tabs. For the evaluation of Cr(VI) removal, the O_2 dosage to the influent of filter bed 1 (1.0, 4.0, 8.0 mg L^{-1}), nitrate concentration (50 mg L^{-1}), and mixture of Cr(VI)-containing raw water with Fe(II)-containing groundwater (1:2 and 1:4) were varied. O_2 dosage variations were first calculated according to the volumetric flow (250 L h^{-1}) and the necessary mass flow of O_2 (setup with mass flow controller 1) to reach the required O_2 concentration. The resulting concentration was measured and verified with an optical dissolved O_2 sensor (FDO 925, WTW—Xylem Analytics Germany Sales GmbH & Co. KG, Weilheim, Germany). For Cr(VI) removal experiments with a nitrate background, Cr(VI) was dosed together with nitrate at dosing point 1, and Fe(II) was subsequently dosed at dosing point 3 (Figure S1, Supplementary Material). For mixing experiments, oxygen-free drinking water was chosen as raw water with a pressure of 2.0 bar and was spiked with Cr(VI) at dosing point 1 to reach the target concentration of 12 μg L^{-1} after mixing the water. At dosing point 3, Fe(II)-containing groundwater under 2.3 bar was mixed in the appropriate ratio (1:2 or 1:4). This way, the sampling of c0 for both Fe(II) and Cr(VI) was possible.

To evaluate the formation of Cr(VI) in the second filter bed, Cr(III) was dosed before entering filter bed 2, and different O_2 doses were applied there (3.0 and 7.5 mL min^{-1}). Reduced groundwater with dissolved Fe(II) and Mn(II) was used as raw water. The O_2 dosage before filter bed 1 was adjusted to the stochiometric O_2 demand for complete Fe(II) oxidation. In filter bed 1, O_2 and Fe(II) (verification by Fe(II) measurements) were completely consumed, while Mn(II) (0.5 mg L^{-1}) was still present in the water.

As standard test setup for all experiments, Cr (12 μg L^{-1}), subsequently Fe(II) (3 mg L^{-1}), and then O_2 (adjusted to the set ORP value of 30 mV after filter bed 1 or with manual control) were dosed into the raw water before entering the relevant filter bed (Figure S1, Supplementary Material). Dosages were controlled with solenoid-driven metering pumps (gamma/X, Prominent; Heidelberg, Germany).

2.4. Analyses

Cr_{tot} samples were directly filtered with membranes (0.45 μm pore size, polyethylene terephthalate, Chromafil, Macherey-Nagel; Düren, Germany) into 50 mL centrifuge tubes (polypropylene (PP)) and stabilized with 300 μL nitric acid (65%). Cr_{tot} was analyzed with a graphite furnace atomic absorption spectroscope (Perkin Elmer 4100ZL, Waltham, MA, United States) with a hollow-cathode lamp (Lumina, Perkin Elmer) at a wavelength of 357.8 nm. The limit of quantification (LOQ) was 2.8 μg L^{-1}, which was determined according to a standardized method, DIN 32645, 2008.

Cr(VI) samples were also directly filtered with membranes (0.45 μm pore size, polyethylene terephthalate, Chromafil, Macherey-Nagel; Düren, Germany) and stabilized by the addition of a 1.5 mL hydrogen phosphate buffer and 450 μL 5 M NaOH into 120 mL bottles (PP) prior to sampling [30]. The samples were again filtered with membranes (0.45 μm pore size, regenerated cellulose, Sartorius; Goettingen, Germany) into 50 mL PP tubes and diluted in a ratio of 1:10 before analyses. Cr(VI) was analyzed with ion chromatography (881 Compact IC Pro, Metrohm; Herisau, Switzerland) with a post-column reaction (Dosino 800) and photometric detector (UV/VIS detector 819, wavelength 530 nm) in accordance with the standardized measurement method US EPA method 218.7. A separation column, A SUPP 5 150/4 (Metrohm; Herisau, Switzerland), was used, and the mobile phase consisted of sodium sulfate/sodium hydroxide solution (15 mmol L^{-1}/0.5 mmol L^{-1}). As a post-column reagent, 2 mmol L^{-1} diphenylcarbazide and 5% methanol with 0.5 mol L^{-1} H_2SO_4 were used. The LOQ was 0.02 μg L^{-1} with a measurement uncertainty of 40% according to a standardized method, DIN 32645, 2008.

Cr(III) concentrations were calculated by subtracting the Cr(VI) concentrations from the Cr_{tot} concentrations.

Fe(II) samples were directly filtered with membranes (0.45 μm pore size, polyethylene terephthalate, Chromafil, Macherey-Nagel; Düren, Germany) into 10 mL glass tubes and immediately analyzed with a UV/VIS photometer (Lambda 35, Perkin Elmer; Waltham, MA, United States) at a wavelength of 510 nm using a triazine-based iron quantification kit (Spectroquant, Merck; Darmstadt, Germany) and a 1 cm quartz cuvette.

Dissolved organic carbon (DOC) was measured with the TOC analysator varioTOC Cube (Elementar Analysensysteme GmbH, Langenselbold, Germany). Samples were directly filtered with membranes (0.45 μm pore size, polyethylene terephthalate, Chromafil, Macherey-Nagel; Düren, Germany), and before measurement, 10 mL of samples were acidified with 80 μL 10% HCl. DOC was fractioned with liquid size-exclusion chromatography with organic carbon detection (LC-OCD) (DOC-Labor GmbH, Karlsruhe, Germany) using a HW50S column (Toyopearl, Japan) according to Huber et al. [31] (2011).

3. Results

3.1. Potential Influences of Oxidizing Water Constituents

First, any critical oxidation agents that potentially could influence the reduction of Cr(VI) with Fe(II) in the supernatant of the filter were identified, and their stabilities were determined by thermodynamic calculations. The lines in Figure 1 display the equimolar distribution of redox pairs (e.g., NO_3^- and NH_4^+), while in the area above the line, the oxidized species dominate, and in the area below, the reduced species dominate. Naturally, only redox reactions are thermodynamically possible when the oxidized species of a redox pair of a higher line reacts with the reduced species of a subjacent line under the formation of the respective redox-pair partners (e.g., Fe(II) reduces O_2 under the formation of $Fe(OH)_3$ and H_2O).

Figure 1 shows that Fe(II) could serve as reducing agent for Cr(VI) over the considered pH range (yellow area) and also that MnO_2 could act as oxidizing agent for Cr(III) for pH values above 3.0 (green area). As Bartlett (1991) [2] and Kotaś and Stasicka (2000) [32] described, MnO_2 constitutes a mediator for Cr(III) oxidation with O_2 in the natural Cr cycle. O_2 and NO_3^- (for pH above 6.0) can act as oxidizing agents towards Cr(III) and therefore could potentially re-oxidize Cr(III) in the RCbF process in the supernatant of filter bed 1 (areas between O_2/H_2O and Cr(III)/Cr(VI) and NO_3^-/NH_4^+ and Cr(III)/Cr(VI)). Furthermore, O_2 and NO_3^- could act as oxidizing agents for Fe(II) over the entire displayed pH range and could therefore negatively influence the targeted Cr(VI) reduction (areas between O_2/H_2O and Fe(II)/Fe(III) and NO_3^-/NH_4^+ and Fe(II)/Fe(III)).

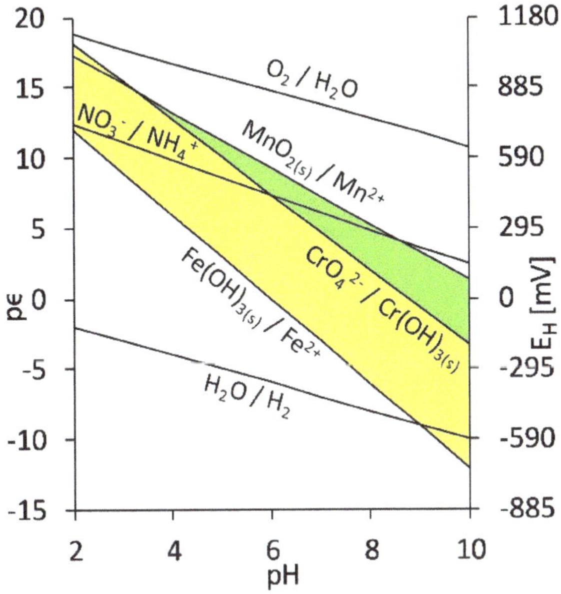

Figure 1. pe pH diagram for relevant redox partners (for 3 mg L^{-1} Fe(II), 0.5 mg L^{-1} Mn(II) and 20 μg L^{-1} Cr(VI)). Yellow area marks pe pH range where Cr(VI) might be reduced by Fe(II) and green area marks pe pH range where Cr(III) might be oxidized by MnO_2.

In the RCbF process, firstly, Cr(VI) is reduced by Fe(II) (Equation (1)), and secondly, the just-formed Cr(III) is coagulated together with an excess of Fe(III) to a Cr(III)–Fe(III) solid solution (Equation (2)) in the supernatant of the filter, which is thirdly filtered out in the subsequent biotic filter bed [27].

$$CrO_4^{2-} + 3Fe^{2+} + 5H_2O \rightleftarrows Cr^{3+} + 3Fe(OH)_{3(s)} + H^+ \quad (1)$$

$$xCr^{3+} + (1-x)Fe^{3+} + 3H_2O \rightleftarrows Cr_xFe_{(1-x)}(OH)_{3(s)} + 3H^+ \ldots (0 < x < 1) \quad (2)$$

3.2. O_2 or NO_3^- as Competing Oxidants and Potential Drivers for Re-Oxidation of Cr(III)

O_2 is the strongest oxidizing agent in Figure 1, indicated as the highest line in the stability field of water between O_2/H_2O and H_2O/H_2, if O_2 functions in a reaction of four electron transfers [33]. However, O_2 in water, as a covalent homodimer, is hindered from reacting with reduced species due to its strong π-double bond and requires a comparably high activation energy to oxidize. Most likely, the oxidation of Fe(II) with O_2 requires four sequential steps (four electron transmissions), whereas O_2 reacts with $Fe(OH)_{2(aq)}$.($Fe(OH)_{2(aq)}$ represents the Fe(II) species (compared to Fe^{2+} and $FeOH^+$) that O_2 oxidizes the fastest [34,35]. The first step (Equation (3)) represents the slowest reaction (one electron transfer) and therefore the kinetics determining this step are as follows [33]:

$$Fe(OH)_{2(aq)} + O_2 \rightleftharpoons Fe(OH)_2^+ + O_2^{-*} \quad (3)$$

To investigate the O_2 dependency on Cr(VI) removal with RCbF, the initial Cr(VI) concentration was set at 12 μg L^{-1} and the Fe(II) dosage was 3 mg L^{-1} at a filtration velocity of 15 m h^{-1}. O_2 reacted poorly with Fe(II) in the supernatant of the filter bed, as shown in Figure 2, but quickly oxidized Fe(II) within the filter bed. As described in the literature, the precipitates on the filter material functioned as an adsorbent for Fe(II) ions and dissolved O_2 and as a catalysator for the required electron transfers due to their semiconductor properties [28,36]. These assumptions explain the occurrence of Fe(II) in the supernatant of the filter bed even though dissolved O_2 is present there, as well as the fast oxidation of Fe(II) and subsequent precipitation in the filter bed (Figure 2). Therefore, in the supernatant of filter 1, two oxidants (Cr(VI) and O_2) competed for the electrons of Fe(II), and after filter bed 1 was reached, no more Fe(II) was present for Cr(VI) reduction. Thus, for an effective RCbF process, the competitive advantage should be attributed to Cr(VI) as an oxidant via the setup of favorable boundary conditions.

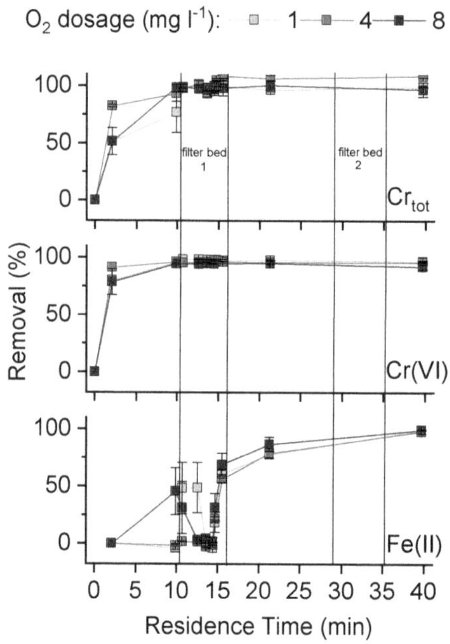

Figure 2. Cumulated removals of Cr_{tot}, Cr(VI), and Fe(II) for different O_2 dosages along the treatment path (given as residence time) at a filtration velocity of 15 m h^{-1}, initial Cr(VI) concentration of 12 µg L^{-1}, Fe(II) dosage of 3.0 mg L^{-1}, pH of 7.0, and a test duration of 50 h. Error bars indicate the maximum and minimum values.

While O_2 is needed in the RCbF process to oxidize excess Fe(II), Figure 2 shows that O_2 did not re-oxidize Cr(III), independent of the O_2 dosage. The removal of Cr was independent of the O_2 dosage, although 8 mg L^{-1} O_2 is, by a stochiometric factor of 19, higher than the amount needed for complete Fe(II) oxidation. In the supernatant of the filter bed, O_2 did not compete with Cr(VI) for the electrons of Fe(II), because the Fe(II) concentration remained almost constant until the filter bed entry. Directly before the filter bed, the Fe(II) concentrations seem to scatter, but this effect was also independent of the O_2 dosage and might have been caused by turbulences at the top layer. The π-double bond probably prevented the dissolved O_2 from exploiting its oxidation potential in the supernatant, and hence Cr(VI) could have been favorably reduced by Fe(II). When the filter bed, with its catalytic surface and microbial activity, was reached, O_2 immediately oxidized Fe(II) and thus did not interfere with the Cr removal. The pH value plays a crucial role in the favored Cr(VI) reduction as opposed to the O_2 reduction in the supernatant. For a pH > 10, the competitive advantage of the electrons changes, and O_2 would be reduced more quickly by Fe(II) compared to Cr(VI). Additionally, the stoichiometric demand for Cr(VI) reduction by Fe(II) would increase (>3 moles Fe(II) per mole Cr(VI)) [37,38]. So in the drinking water pH range, the advantage of Cr(VI) to function as an oxidant is used in the filter supernatant. As for the oxidation of Fe(II) with O_2, the speciation of Fe(II) was described as crucial for the oxidation of Fe(II) with Cr(VI). As for the oxidation with O_2, non-valent $Fe(OH)_{2(aq)}$ was postulated as being the Fe(II) species that reduces Cr(VI) the fastest [37]. As described by Sigg and Stumm (2016) [33], the unhydrolyzed species Fe^{2+} is a weaker reductant because of reduced electron density of the Fe(II) atom. Hence, the valence and electron density, in combination with the hydrolyzation of the Fe(II) species, are crucial for the redox activity of Fe(II). On the other hand, Cr(VI) occurs as a two-valent chromate anion in the drinking water pH range, and its valence was identified as a negligible factor for the reduction kinetics with Fe(II) [37].

NO_3^- is regulated with a limit value of 50 mg L^{-1} (according to the EU Drinking Water Directive [19]) and could provide an oxidant concentration that is by a factor of 5000 higher than the maximal observed Cr concentration in Germany (i.e., in groundwater [39]). NO_3^- has been described as a potent oxidant in groundwater, e.g., for pyrite, that can mobilize heavy metals that might reach drinking water [40]. Since Cr is more mobile as oxidized species (Cr(VI)), the inputs of NO_3^- into aquifers might increase the threat of Cr occurrence. Thus, the co-existence of NO_3^- and Cr(VI) might have negative consequences for Cr(VI) removal in the RCbF process. However, as Figure 3 shows, NO_3^- did not negatively influence the RCbF process. The experimental setup was again 12 µg L^{-1} as the initial Cr(VI) concentration, a Fe(II) dosage of 3.0 mg L^{-1}, and a filtration velocity of 15 m h^{-1}. For pH values higher than 6.0, NO_3^- could potentially re-oxidize Cr(III) to Cr(VI) and compete with Cr(VI) as an oxidant and oxidize Fe(II) to Fe(III) (Figure 1). However, an oxidation of Fe(II) in the supernatant of the filter bed was not observed, and hence, Fe(II) could be utilized to its full extend as a reductant for Cr(VI) reduction. Although the pH was 7.0, no re-oxidation of Cr(III) by NO_3^- could be observed. Cr(III) removal lagged behind Cr(VI) removal in the supernatant of the filter, and consequently, not the Cr(VI) reduction, but the formation of the chromium(III)–iron(III) hydroxide solid solution seemed to be the kinetic determining step in the RCbF process (Figure 3).

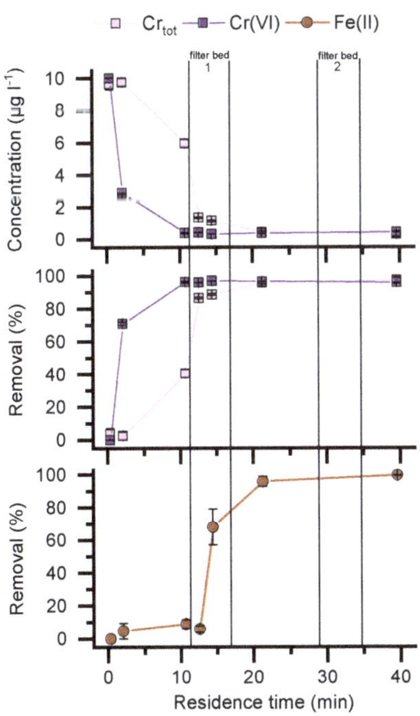

Figure 3. Cr(VI), Cr_{tot}, and Fe(II) removals after nitrate dosage (50 mg L^{-1}) over residence time. Filtration velocity 15 m h^{-1}, initial Cr(VI) concentration of 12 µg L^{-1}, Fe(II) dosage of 3.0 mg L^{-1}, pH of 7.0, and test duration of 50 h. Error bars indicate the maximum and minimum values.

3.3. Oxidation of Cr(III) to Cr(VI) by O_2 with MnO_2 as a Mediator

In the natural Cr cycle, Cr(III) is oxidized to mobile Cr(VI) on MnO_2, which serves as a mediator between O_2 and Cr(III) in soil (Equation (4)) [2].

$$4\,Cr(OH)_{3(aq)} + 3\,O_2 \overrightarrow{MnO_2}\, 4\,CrO_{42-} + 8\,H^+ + 2\,H_2O \quad (4)$$

Cr(III), with a much lower solubility than Cr(VI), can be mobilized by either a shift in the pH to below 5.0 or by a complexation with natural organic matter [2,32]. The pH very seldom decreases below 5.0 in drinking water treatment, but a complexation by DOC seems a likely mechanism for Cr(III) mobilization. The DOC of the test water was 3.0 mg L^{-1} and showed pronounced fractions of humic acids, which were described to influence the solubility of Cr(III) [32]. The DOC concentrations and composition were not influenced by the treatment process, as is displayed in Figure S2, Supplementary Material. However, no detailed examination of how DOC influences the mobilization of Cr in the context of water treatment is known to the authors, and thus further research on the topic is required. Furthermore, the solubility of Cr(III) at a neutral pH is approx. 4 µg L^{-1} [41], and raw water with dissolved Mn(II) and complexed Cr(III) is chemically stable and thus likely to occur in the environment (Figure 1). The initial Cr(III) concentration was 12 µg L^{-1} at a filtration velocity of 15 m h^{-1}. The Cr(III) was quickly removed in filter bed 2 with biotically generated MnO_2, but Cr(III) was partially oxidized to Cr(VI), which broke through the filter, as shown Figure 4. The ratio of oxidized Cr(VI) increased with an increasing O_2 dosage, and the highest measured concentrations of Cr(VI) in the effluent corresponded with 1/3 of the initial Cr(III) concentration. Thus, the oxidation of Cr(III) to Cr(VI) proceeded very quickly in only a 5 min contact time in the filter bed. Thus, Cr(III) must be fully retained in the filter bed for iron removal to avoid its re-oxidation with O_2 with MnO_2 as a mediator. Most likely, mechanistically, the O_2 was adsorbed onto the MnO_2 surface, and Cr(III) was enriched in the filter, and thus, MnO_2 could catalytically promote the oxidation of Cr(III) with O_2.

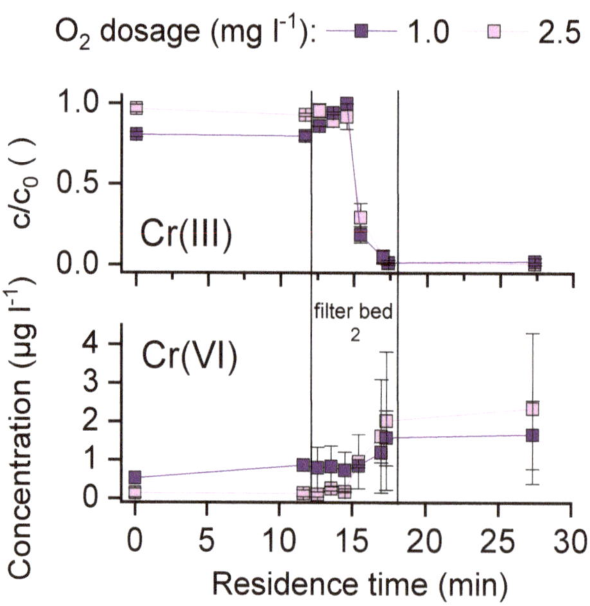

Figure 4. Cr(III) and Cr(VI) concentrations for different O_2 dosages over residence times in the second filter (12 µg L^{-1} Cr(III), with a filtration velocity of 15 m h^{-1}, a pH of 7.0, and 0.5 mg L^{-1} Mn(II)). Error bars indicate the maximum and minimum values during the 50 h operation.

3.4. Mixing of Cr(VI) Containing Water with Fe(II) Containing Groundwater

The continuous chemical consumption due to the dosage of Fe(II) is a disadvantage of the RCbF process. Therefore, the possibility of mixing anoxic Fe(II) containing groundwater with oxic Cr(VI) containing raw water was assessed as a chemical-free treatment. Of course, only if these types of water are locally available is this concept is possible. The experimental setup was an initial Cr(VI) concentration of 12 μg L^{-1} after mixing at a filtration velocity of 15 m h^{-1}. The removal of Cr was made successful by mixing the different waters; however, the Cr removal was dependent on the resulting Fe(II) concentration after mixing (Figure 5), as expected. While a mixing ratio of 1:2 led to a sufficient removal with an effective Fe(II) concentration of 1.2 mg L^{-1}, a ratio of 1:4 led to an insufficient Cr(VI) removal due to a resulting Fe(II) concentration of only 0.6 mg L^{-1}. These results are in line with previous results that identified Fe(II) dosage as a key factor [27].

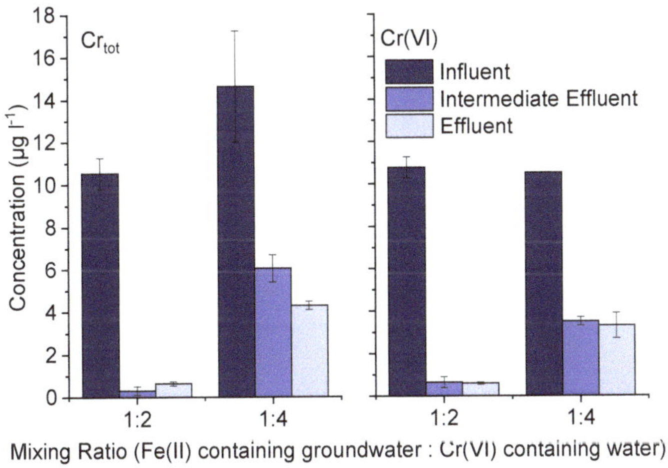

Figure 5. Cr(VI) and Cr$_{tot}$ concentrations after mixing of Cr(VI) spiked water (Cr(VI) concentration after mixing was 12 μg L^{-1}) with Fe(II) containing anoxic groundwater (Fe(II) = 2.4 mg L^{-1}) in different mixing ratios over residence time (filtration velocity: 15 m h^{-1}; pH: 7.0). Error bars indicate the maximum and minimum values during the 50 h operation.

4. Conclusions

The redox pair Cr(VI) and Cr(III) is very reactive, and different reducing and oxidizing agents might affect the oxidation state of Cr and hence its mobility. O_2 (independent of pH), NO_3^- (for pH above 6.0) or MnO_2 (for pH below 9.0) could potentially oxidize Cr(III) to Cr(VI) and thus increase the mobility of Cr. Fe(II) (independent of pH) can reduce Cr(VI) to Cr(III), which consequently precipitates and can be removed via filtration in the RCbF process. If the oxidants O_2 and NO_3^- are present, however, the Cr(VI) reduction by Fe(II) might be affected due to competing oxidation. However, O_2 and NO_3^- did not negatively affect the RCbF process. Neither the oxidation of Fe(II) by O_2 or NO_3^- nor the re-oxidation of just-produced Cr(III) were observed in the supernatant of the filter bed. The oxidation of Cr(III) to Cr(VI) in filter bed 2 happened quickly, dependent on the O_2 dosage. MnO_2 served as a mediator for Cr(III) oxidation with O_2, most likely due to an enrichment of Cr(III) on the filter bed surface and an adsorption of O_2, so that MnO_2 could catalytically promote oxidation. While the Cr(VI) reduction to Cr(III) was completed in the supernatant of the filter bed, the formation of filterable Cr(III) hydroxide happened more slowly, and hence it was identified as a kinetic-determining step for Cr removal with RCbF. If the local conditions allow, a chemical-free variation of the RCbF process is possible by mixing Cr(VI) containing oxic raw water with Fe(II) containing anoxic groundwater; however, this is dependent on the resulting Fe(II) concentration.

Supplementary Materials: The following supporting information can be downloaded at https://www.mdpi.com/article/10.3390/w15193363/s1: Figure S1: Simplified scheme of the pilot plant at technical scale; Table S1: Composition of drinking water (groundwater), which was used as test water; Figure S2: DOC development in pilot plant; left: total DOC; right: fractioned DOC with LC-OCD.

Author Contributions: D.M., conceptualization; A.S.R., S.S.Z. and D.M., data curation; S.S.Z. and D.M., formal analysis; D.M., funding acquisition; S.S.Z. and D.M., investigation; D.M. and S.S.Z., Methodology; D.M., project administration; D.M., resources; D.M., software; D.M., supervision; D.M. and A.S.R., validation; D.M., visualization; D.M. and A.S.R., roles/writing—original draft; D.M., writing—review and editing. All authors have read and agreed to the published version of the manuscript.

Funding: Project funding from federal funds in the area of the Federal Ministry of Health; Funding Code: Chapter 1501 Title 54401 (FKZ 2517FSB452).

Data Availability Statement: The original contributions presented in the study are included in the article. Further inquiries can be directed to the corresponding author.

Acknowledgments: We gratefully acknowledge the support and funding from the Federal Ministry of Health. We also thank UBA colleagues in Berlin-Marienfelde, especially Silke Pabst for her support and great analytical work, Andrea Steuer for her language corrections, and Robert Eschrich for his engineering support.

Conflicts of Interest: The authors declare that they have no known competing financial interest or personal relationships that could have appeared to influence the work reported in this paper.

Abbreviations

Cr: chromium, Fe(II): divalent iron, DOC: dissolved organic carbon, LOQ: limit of quantification, Cr(VI): hexavalent chromium, PP: polypropylene, RCF: reduction, coagulation, and filtration, RCbF: reduction, coagulation, and biotic filtration, Cr_{tot}: total chromium, Cr(III): trivalent chromium, UBA: German Environment Agency, ORP: oxidation/reduction potential, WHO: World Health Organization.

References

1. Barnhart, J. Occurrences, Uses, and Properties of Chromium. *Regul. Toxicol. Pharmacol.* **1997**, *26*, S3–S7. [CrossRef]
2. Bartlett, R.J. Chromium Cycling in Soils and Water: Links, Gaps, and Methods. *Env. Health Perspect* **1991**, *92*, 17–24. [CrossRef]
3. Rai, D.; Eary, L.E.; Zachara, J.M. Environmental Chemistry of Chromium. *Sci. Total Environ.* **1989**, *86*, 15–23. [CrossRef] [PubMed]
4. Rai, D.; Sass, B.M.; Moore, D.A. Chromium(III) Hydrolysis Constants and Solubility of Chromium(III) Hydroxide. *Inorg. Chem.* **1987**, *26*, 345–349. [CrossRef]
5. Sass, B.M.; Rai, D. Solubility of Amorphous Chromium(III)-Iron(III) Hydroxide Solid Solutions. *Inorg. Chem.* **1987**, *26*, 2228–2232. [CrossRef]
6. McCarroll, N.; Keshava, N.; Chen, J.; Akerman, G.; Kligerman, A.; Rinde, E. An Evaluation of the Mode of Action Framework for Mutagenic Carcinogens Case Study II: Chromium (VI). *Environ. Mol. Mutagen.* **2010**, *51*, 89–111. [CrossRef]
7. Sedman, R.M.; Beaumont, J.; Mcdonald, T.A.; Reynolds, S.; Krowech, G.; Howd, R. Review of the Evidence Regarding the Carcinogenicity of Hexavalent Chromium in Drinking Water. *J. Environ. Sci. Health* **2006**, *24*, 155–182. [CrossRef]
8. Stern, A.H. A Quantitative Assessment of the Carcinogenicity of Hexavalent Chromium by the Oral Route and Its Relevance to Human Exposure. *Environ. Res.* **2010**, *110*, 798–807. [CrossRef]
9. Zhitkovich, A. Chromium in Drinking Water: Sources, Metabolism, and Cancer Risks. *Chem. Res. Toxicol.* **2011**, *24*, 1617–1629. [CrossRef]
10. De Flora, S. Threshold Mechanisms and Site Specificity in Chromium(VI) Carcinogenesis. *Carcinogenesis* **2000**, *21*, 533–541. [CrossRef]
11. Kopec, A.K.; Kim, S.; Forgacs, A.L.; Zacharewski, T.R.; Proctor, D.M.; Harris, M.A.; Haws, L.C.; Thompson, C.M. Genome-Wide Gene Expression Effects in B6C3F1 Mouse Intestinal Epithelia Following 7 and 90days of Exposure to Hexavalent Chromium in Drinking Water. *Toxicol. Appl. Pharmacol.* **2012**, *259*, 13–26. [CrossRef]
12. Moffat, I.; Martinova, N.; Seidel, C.; Thompson, C.M. Hexavalent Chromium in Drinking Water: JOURNAL AWWA. *J.—Am. Water Work. Assoc.* **2018**, *110*, E22–E35. [CrossRef]
13. Thompson, C.M.; Haws, L.C.; Harris, M.A.; Gatto, N.M.; Proctor, D.M. Application of the U.S. EPA Mode of Action Framework for Purposes of Guiding Future Research: A Case Study Involving the Oral Carcinogenicity of Hexavalent Chromium. *Toxicol. Sci.* **2011**, *119*, 20–40. [CrossRef]

14. Thompson, C.M.; Kirman, C.R.; Proctor, D.M.; Haws, L.C.; Suh, M.; Hays, S.M.; Hixon, J.G.; Harris, M.A. A Chronic Oral Reference Dose for Hexavalent Chromium-Induced Intestinal Cancer: Reference Dose for Cr(VI)-Induced Intestinal Cancer. *J. Appl. Toxicol.* **2014**, *34*, 525–536. [CrossRef]
15. Thompson, C.M.; Rager, J.E.; Suh, M.; Ring, C.L.; Proctor, D.M.; Haws, L.C.; Fry, R.C.; Harris, M.A. Transcriptomic Responses in the Oral Cavity of F344 Rats and B6C3F1 Mice Following Exposure to Cr(VI): Implications for Risk Assessment: Transcriptomic Responses in the Oral Cavity. *Environ. Mol. Mutagen.* **2016**, *57*, 706–716. [CrossRef]
16. National Toxicology Program. Toxicology and Carcinogenesis Studies of Sodium Dichromate Dihydrate (Cas No. 7789-12-0) in F344/N Rats and B6C3F1 Mice (Drinking Water Studies). *Natl. Toxicol. Program Tech. Rep. Ser.* **2008**, *546*, 1–192.
17. WHO. *Chromium in Drinking-Water—Background Document for Development of WHO Guidelines for Drinking-Water Quality*; World Health Organization: Geneva, Switzerland, 2020; p. 38.
18. Roller, M. *Potentielle Schädlichkeit von Chrom Im Trinkwasser—Einordnung Der Epidemiologischen Befunde Zum Krebsrisiko Nach Exposition von Populationen Gegenüber Chrom(VI) Im Trinkwasser Und Vorschlag Zur Ableitung Einer Expositions-Risikobeziehung*; Beratungsbüro für Risikoabschätzung: Dortmund, Germany, 2012; p. 79.
19. EU. Directive (EU) 2020/2184 of the European Parliament and of the Council of 16 December 2020 on the Quality of Water Intended for Human Consumption. 2020, p. 62. Available online: https://environment.ec.europa.eu/topics/water/drinking-water_en (accessed on 21 September 2023).
20. BMG. German Drinking Water Ordinance. 2023. Available online: https://www.gesetze-im-internet.de/trinkwv_2023/TrinkwV.pdf (accessed on 21 September 2023).
21. California Water Boards. *State Water Resources Control Board Proposed Hexavalent Chromium MCL Staff Report*; California Department of Public Health: Sacramento, CA, US, 2022; p. 9.
22. Blute, N.; Wu, X.; Cron, C.; Abueg, R.; Froelich, D.; Fong, L. Hexavalent Chromium Treatment Implementation in Glendale, Calif. *J.—Am. Water Work. Assoc.* **2014**, *106*, E160–E175. [CrossRef]
23. Allendorf, A.; Bauer, K.-H.; Huschens, O.; Post, B.; Weygand, A.; Riegel, M.; Schlitt, V.; Sacher, F. *Untersuchungen Zum Vorkommen von Sechswertigem Chrom Und Seiner Entfernung Bei Der Trinkwasseraufbereitung*; Deutsche Verein des Gas- und Wasserfachs: Bonn, Germany, 2016; p. 149.
24. Lee, G.; Hering, J.G. Removal of Chromium(VI) from Drinking Water by Redox-Assisted Coagulation with Iron(II). *J. Water Supply: Res. Technol.—Aqua* **2003**, *52*, 319–332. [CrossRef]
25. McGuire, M.J.; Blute, N.K.; Seidel, C.; Qin, G.; Fong, L. Pilot-Scale Studies of Hexavalent Chromium Removal from Drinking Water. *J.—Am. Water Work. Assoc.* **2006**, *98*, 134–143. [CrossRef]
26. Sharma, S.K.; Branislav, P.; Gary, A. Chromium Removal from Water: A Review. *J. Water Supply Res. Technol.-Aqua* **2008**, *57*, 541–553. [CrossRef]
27. Mahringer, D.; Zerelli, S.S.; Dippon, U.; Ruhl, A.S. Pilot Scale Hexavalent Chromium Removal with Reduction, Coagulation, Filtration and Biological Iron Oxidation. *Sep. Purif. Technol.* **2020**, *253*, 117478. [CrossRef]
28. DVGW. *Enteisenung Und Entmanganung—Teil 1: Grundsätze Und Verfahren*; Deutsche Verein des Gas- und Wasserfachs: Bonn, Germany, 2005; p. 34.
29. Mahringer, D.; Zerelli, S.S.; Ruhl, A.S. Pilot-Scale Vanadium Adsorption onto in Situ Biogenic Amorphous Ferric Hydroxide. *J. Water Process Eng.* **2023**, *56*, 104278. [CrossRef]
30. Mahringer, D.; Polenz, C.; El-Athman, F. Stabilization of Chromium (VI) in the Presence of Iron (II): Method Development and Validation. *Water* **2020**, *12*, 924. [CrossRef]
31. Huber, S.A.; Balz, A.; Abert, M.; Pronk, W. Characterisation of Aquatic Humic and Non-Humic Matter with Size-Exclusion Chromatography—Organic Carbon Detection—Organic Nitrogen Detection (LC-OCD-OND). *Water Res.* **2011**, *45*, 879–885. [CrossRef]
32. Kotaś, J.; Stasicka, Z. Chromium Occurrence in the Environment and Methods of Its Speciation. *Environ. Pollut.* **2000**, *107*, 263–283. [CrossRef] [PubMed]
33. Sigg, L.; Stumm, W. *Aquatische Chemie: Einführung in Die Chemie Natürlicher Gewässer*, 6th ed.; UTB Naturwissenschaften, Umweltwissenschaften; 6., vollst. überarb. Aufl.; vdf Hochschul-Verl: Zürich, Switzerland, 2016; ISBN 978-3-7281-3768-5.
34. Millero, F.J. The Effect of Ionic Interactions on the Oxidation of Metals in Natural Waters. *Geochim. Et Cosmochim. Acta* **1985**, *49*, 547–553. [CrossRef]
35. Morgan, B.; Lahav, O. The Effect of PH on the Kinetics of Spontaneous Fe(II) Oxidation by O2 in Aqueous Solution—Basic Principles and a Simple Heuristic Description. *Chemosphere* **2007**, *68*, 2080–2084. [CrossRef]
36. Grothe, P.; Czekalla, C. Enteisenung und Entmanganung. In *Wasseraufbereitung—Grundlagen und Verfahren*; Gimbel, R., Jekel, M., Deutsche Vereinigung des Gas- und Wasserfaches, Eds.; Lehr- und Handbuch Wasserversorgung; Oldenbourg Industrieverl: München, Germany, 2004; ISBN 978-3-486-26365-7.
37. Buerge, I.J.; Hug, S.J. Kinetics and PH Dependence of Chromium (VI) Reduction by Iron (II). *Environ. Sci. Technol.* **1997**, *31*, 1426–1432. [CrossRef]
38. Eary, L.E.; Rai, D. Chromate Removal from Aqueous Wastes by Reduction with Ferrous Ion. *Environ. Sci. Technol.* **1988**, *22*, 972–977. [CrossRef]

39. BGR. *SGD WMS Information Der BGR Hannover: Hydrogeologische Karte von Deutschland 1:200.000, Hintergrundwerte (HÜK200 HGW); Thema Hydrogeochemie, Hintergrundwerte Im Grundwasser*; BGR: Hannover, Germany, 2014.
40. van Berk, W.; Fu, Y. Redox Roll-Front Mobilization of Geogenic Uranium by Nitrate Input into Aquifers: Risks for Groundwater Resources. *Environ. Sci. Technol.* **2017**, *51*, 337–345. [CrossRef]
41. Rai, D.; Moore, D.A.; Hess, N.J.; Rosso, K.M.; Rao, L.; Heald, S.M. Chromium (III) Hydroxide Solubility in the Aqueous K^+-H^+-OH^--CO_2-HCO_3^--CO_3^{2-}-H_2O System: A Thermodynamic Model. *J. Solut. Chem.* **2007**, *36*, 1261–1285. [CrossRef]

water

Review

Sustainable Utilization of Pulp and Paper Wastewater

Xiaoli Liang [1,2], Yanpeng Xu [1,2], Liang Yin [3], Ruiming Wang [1,2], Piwu Li [1,2], Junqing Wang [1,2] and Kaiquan Liu [1,2,*]

1 State Key Laboratory of Biobased Material and Green Papermaking (LBMP), Qilu University of Technology (Shandong Academy of Sciences), Jinan 250353, China; 10431211135@stu.qlu.edu.cn (X.L.); 10431211110@stu.qlu.edu.cn (Y.X.); wrm@qlu.edu.cn (R.W.); piwuli@qlu.edu.cn (P.L.); wangjq@qlu.edu.cn (J.W.)

2 Key Laboratory of Shandong Microbial Engineering, College of Bioengineering, Qilu University of Technology (Shandong Academy of Sciences), Jinan 250353, China

3 Laboratory of Microalgae Genetic Engineering, Gansu Engineering Technology Research Center for Microalgae, Hexi University, Zhangye 734000, China; yinl03@163.com

* Correspondence: liukq@qlu.edu.cn

Abstract: The pulp and paper industry plays an important role in the global economy and is inextricably linked to human life. Due to its large scale, the production process generates a large amount of wastewater, which poses a major threat to the environment. The sustainable utilization and safe treatment of pulp and paper wastewater can effectively reduce environmental pollution, improve resource utilization efficiency, protect water resources, provide economic benefits for pulp and paper enterprises, and thus promote the green and sustainable development of the pulp and paper industry. Therefore, this study discusses the pollution components of pulp and paper wastewater and their impact on the environment and human health. In this review, we aim to explore the sustainable development of pulp and paper wastewater by summarizing the characteristics of current pulp and paper wastewater, the commonly used treatment methods for pulp and paper wastewater, the application of pulp and paper wastewater recycling, and the future development direction of pulp and paper wastewater.

Keywords: pulp; paper; wastewater treatment; sustainable utilization

1. Introduction

Pulp and paper is an industrial process in which finely divided raw materials such as wood, straw, or other plant fibers are used to produce paper via various processes such as crushing, soaking, grinding, fiber separation, and coagulation [1]. The pulp and paper industry is one of the most important industrial sectors in the world, and its global output value in 2020 reached USD 580 billion. People's lives are closely related to the pulp and paper industry. Global apparent consumption of paper and paperboard in 2021 is 428.51 million tons, up 6.6% from 401.81 million tons in 2020. Global apparent consumption per capita was 55.1 kg. North America has the highest apparent consumption per capita at 197.6 kg, followed by Europe and Oceania at 116.8 kg and 96.0 kg, respectively. Per capita apparent consumption is 48.5 kg in Asia, 45.1 kg in Latin America, and only 7.2 kg in Africa [2]. An overview of the pulp and paper industry capacity in 2022 in seven countries, including Australia, Brazil, and Canada, is shown in Table 1 [3]. According to relevant statistics, the total output of paper and paperboard in the global pulp and paper industry was ~417 million tons in 2022, of which the total output of the United States was ~65.95 million tons and that of China reached ~124.32 million tons [4]. Being the largest developing country in the world, China ranks first in the total amount of paper products in the world; however, its per capita output of paper products is considerably lower than that of the developed countries. Thus, China's pulp and paper industry has great potential for development. According to statistics, China's pulp, paper, and paper products industry achieved a total output of 283.91 million tons of pulp, paper, paperboard, and paper products in 2022, which reflects an annual increase of 1.32%. In that year, the paper and paperboard output was 124.25 million tons, which reflects an increase of 2.64% over the previous year. The pulp output was 85.87 million tons, which reflects an increase of 5.01% over the previous year, and the paper products output was 73.79 million tons, a decrease of 4.65% over the previous year [4]. The industry's operating income was USD 218.7 billion, indicating an increase of 0.44% over the previous year; the total profit was USD 8.9 billion, an annual growth of −29.79%. In 2022, there were ~2500 paper and paperboard manufacturers in China, and the production volume of paper and paperboard in China was 124.25 million tons, a 2.64% increase over the previous year. The consumption was 124.03 million tons, which reflects an increase of −1.94% over the previous year, and the per capita annual consumption was 87.84 kg [4]. From 2013 to 2022, the average annual growth rate of China's paper and paperboard production was 1.87% and that of consumption was 2.59%.

Table 1. Overview of pulp and paper industry capacity in various countries in 2022.

	Total Pulp Production (1000 Tones Air)				
Country	**Wood Pulp for Paper and Paperboard**	**Pulp of Other Fibre for Paper and Paperboard**	**Dissolving Pulp, Wood + Other Raw Materials**	**Paper and Paperboard**	**Utilization of Recovered Paper for Making Paper and Paperboard**
Australia	923	0	0	3024	1691
Brazil	24,969	-	670	11,040	5090
Canada	13,600	-	-	8600	2850
China	21,150	5580	-	124,320	66,420
America	40,822	-	-	65,959	29,054
Japan	7579	4	155	13,661	15,947
Korea	277	-	-	11,254	8315
Thailand	1082	151	89	5374	5201

The development of the pulp and paper industry is an important economic activity. However, wastewater generated during related processes is an important source of environmental pollution [5,6]. Wastewater pollutants produced by the pulp and paper industry include organic and inorganic substances. Organic substances include lignin, starch, amylase, and cellulose, whereas inorganic substances primarily include sulfates, chlorides, and nitrates [7]. In recent years, the treatment and sustainable use of pulp and paper wastewater has become an important environmental concern over the globe [8].

Since the development of the pulp and paper industry, different pulp methods such as mechanical, chemical, chemical–mechanical, and biomechanical pulp have been developed to meet different paper needs. At present, the chemical pulp method is extensively employed in the industry, and primarily includes alkaline, kraft, sulfite, organic solvent, and mechanochemical pulp [9,10]. Chemical pulp involves the separation of lignin from fibrous material to obtain cellulose by exposing papermaking materials to chemicals. This method is highly productive and extensively employed in the paper industry, accounting for ~75% production of the total industrial and commercial paper in the market [11]. However, this method produces a large amount of wastewater. It is also one of the main sources of wastewater in the pulp and paper industry. Pulp and paper wastewater includes wastewater from pulp and from different stages of paper. The composition of wastewater is considerably complex [5,12]. Figure 1 shows the sources of wastewater from the pulp and paper process.

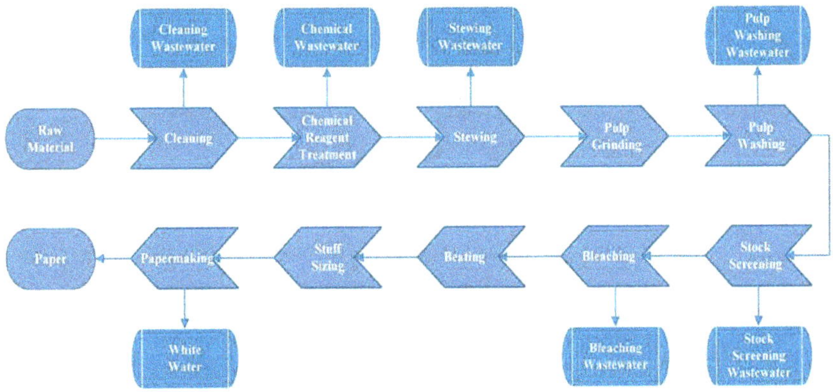

Figure 1. Sources of wastewater from the pulp and paper process.

The pulp process involves disintegrating paper raw materials, separating most of the fibers, separating the remaining pulp by adding various chemical paper agents to ensure good quality of the paper, and bleaching. At this stage, a large amount of high-concentration pulp wastewater is produced [13].

Black liquor is mainly generated from wastewater produced during cooking, which is the most polluting wastewater in the entire paper process. It contains a large amount of organic matter, a high concentration of chemical oxygen demand (COD), and high chroma along with a large amount of refractory fibers and inorganic salts [14]. Wastewater in the middle stage of paper primarily refers to all wastewater except black liquor, including washing, bleaching, and paper wastewater discharged during the paper process [15]. Wastewater from the washing and screening process is similar to black liquor; however, its concentration is not as high as that of black liquor [1,16]. The paper bleaching process separates the remaining fiber pigments. Due to the addition of a large amount of chlorine containing bleaching agent, the discharged wastewater contains a large amount of chlorine-containing organic matter, which has high chroma and toxicity [17]. During the paper process, the pulp is transformed into paper. Wastewater discharged from this process is also called white water and contains various fibers and rubber materials, fillers, and preservatives that were added during the paper process. Compared with black liquor, the pollution caused by white water is minor, i.e., white water is slightly less toxic than bleaching wastewater [18].

Low-molecular-weight compounds (chlorophenols and other organochlorine compounds) in pulp wastewater discharged during the pulp process are the main cause of biomutagenicity and bioaccumulation in rivers due to their hydrophobicity and ability to penetrate cell membrane sources [19]. The toxic effects of absorbable organic halide range from carcinogenicity and mutagenicity to acute toxicity. Their accumulation and slow degradation can lead to high COD. Research has shown that the discharge of pulp and paper wastewater not only increases the toxicity caused to phytoplankton and zooplankton due to increasing biochemical oxygen demand, COD, and other pollution parameters but also induces the growth of fecal coliform bacteria, which poses a health hazard to these organisms [20]. Approximately half of the chemicals that can affect the endocrine system are chlorides (such as dioxins and polychlorinated biphenyls); studies have confirmed that these chlorinated hydrocarbon organic pollutants are primarily produced by their chemical structural changes due to lignin involved in the cooking and bleaching processes [12]. When chemical substances that are similar in structure to biological hormones enter the human body, they destroy the balance of human hormones, resulting in endocrine disorders and, in severe cases, abnormal development and reproductive functions [21]. The US Environmental Protection Agency (EPA) surveyed some paper mills and found that

60% of wastewater discharged from the paper mills contained dioxin components [22]. Dioxin exhibits the characteristics of teratogenicity, carcinogenicity, and mutagenicity [23]. It not only affects the secretion of hormones that feminize males and influence the growth and development of children but can also causes multiple brain neuropathy and acute toxicity [24]. Its toxicity is equivalent to more than 1000 times that of potassium cyanide. As a polymer compound, lignin has some molecular structures similar to the basic structure of dioxins, which are very easy to produce when pulp is bleached using chlorine [25]. Thus, the most effective method for minimizing dioxin emissions from pulp mills is limiting the use of chlorine-containing compounds [26]. Pulp and paper wastewater also contains some nonpolar compounds, such as halogenated hydrocarbons and polychlorinated biphenyls, making their water solubility considerably low [5]. In pulp and paper wastewater, chloroform, pentachlorophenol, chlorinated vanilla aldehydes, chlorinated guaiacol, and other substances exist [27]. Such toxic substances are not easily degraded biochemically or non-biochemically, and their discharge into natural water bodies has a toxic effect on organisms. The substances have strong toxicity and carcinogenic effects; at low concentrations, they chronically accumulate to cause lesions, and at high concentrations, they cause death [28]. They are enriched through the food chain or directly affect mammals and humans through drinking water [29]. Consequently, some countries have established limits for the discharge of absorbable organic halides per ton of the pulp from the paper industry [10].

With the development of society, people now have increased awareness of environmental protection and pay increased attention to sewage generated by various industries. Various sewage treatment methods have been developed and applied accordingly [1]. At present, technologies for the sustainable use of pulp and paper wastewater primarily include sewage treatment, biological treatment, recycling, and comprehensive utilization [30]. Among them, sewage treatment is the most commonly used technology at present; its purpose is to reduce pollutants in wastewater. Organic matter in sewage is primarily treated by biological treatment, whereas inorganic matter is primarily treated via chemical treatment to reduce pollutants [23]. Biological treatment is based on the microbial analysis of organic matter in wastewater. Its advantage is high treatment efficiency, and the disadvantage is high energy consumption [31]. Wastewater reuse technology involves reusing wastewater during the pulp and paper process. The advantage of this technology is that it can effectively reduce the discharge of wastewater; however, its disadvantage is that it requires advanced treatment of organic and inorganic substances in wastewater to meet the discharge standards [32]. Comprehensive utilization involves converting the pollutants in wastewater into useful substances. The advantage of this technology is that it can effectively utilize pollutants in wastewater; the disadvantage is that the related technology is complex and expensive [33].

When paper wastewater is discharged without proper treatment, it can alter the natural water body's environment from aerobic to anaerobic, disrupt the acid–base balance, diminish the water body's self-purification capabilities, and ultimately lead to water quality deterioration. Thus, the effective treatment of pulp and paper wastewater is of utmost importance. Recycling valuable substances within this wastewater and implementing industrial wastewater recycling within or between enterprises can not only safeguard water environments and ecological balance but also conserve water resources, reduce pollutants, and promote sustainable development. This effort has important implications for mitigating water resource shortages, reducing the influx of toxic substances into the ecological environment, minimizing water production costs for businesses, advancing economic and societal development, and attaining "carbon reduction" objectives. Achieving sustainable use of pulp and paper wastewater is a multifaceted challenge that requires comprehensive consideration of technical, economic, and environmental factors to identify optimal solutions [6].

We aim to explore the sustainable development of pulp and paper wastewater by summarizing the characteristics of current pulp and paper wastewater, the commonly used treatment methods for pulp and paper wastewater, the application of pulp and paper wastewater recycling, and the future development direction of pulp and paper wastewater!

2. Literature Search

The scientific documents related to this review are mainly from the CNKI database, the WOS database and the FAOLEX database.

In the process of database retrieval, several combinations of the following keywords are applied: pulp; paper; wastewater treatment; sustainable utilization. All references are original articles published before October 2023. We removed duplicates and selected by title, keyword, summary, relevance. A total of 95 publications covering pulp and paper making, pulp and paper wastewater treatment, and wastewater treatment for further processing and utilization met our inclusion criteria. We acknowledge that some documents may have been omitted; however, we believe that the full range of studies collected faithfully represents the current state of knowledge on the subject.

Our results consist of three main parts. The first part mainly introduces the characteristics of pulp and paper wastewater, this part quotes 15 references. The second part mainly introduces common treatment methods for pulp and paper wastewater, 25 references are cited in this part. In this paper, the advantages and disadvantages of common methods used for treating pulp and paper wastewater are compared by table, and corresponding references are provided. The third part mainly introduces the utilization of pulp and paper wastewater. In this part, 16 references are quoted to introduce the relevant contents of the recycling of pulp and paper wastewater.

3. Characteristics of Pulp and Paper Wastewater

Wastewater generated by the pulp and paper industry is characterized based on chemical pollution and microbial pollution [34].

3.1. Chemical Pollution Characteristics

Pulp and paper wastewater is a high-concentration multicomponent pollutant. Its main constituents include suspended solids, carbohydrates, ammonia nitrogen, cyanide, sulfate, phosphate, heavy metals, and the different concentration ranges are shown in Table 2 [35].

Table 2. Different concentration ranges of chemical pollutants.

Pollutant	Content (mg/L)	Test Method
Suspended solids	>500 mg/L	Gravimetric determination; Direct microscopic count
Carbohydrates	100~500 mg/L	Combustion oxidation nondispersive infrared absorption method
Ammonia nitrogen	50~200 mg/L	UV spectrophotometry
Cyanide	5~20 mg/L	UV spectrophotometry
Sulfate compound	100~500 mg/L	Barium chromate spectrophotometry
Phosphate	>20 mg/L	Ammonium molybdate spectrophotometry

Suspended solids: pulp and paper wastewater contains a large amount of suspended solids, such as pulp debris, cellulose, and lignin. The content of these suspended solids can reach more than 500 mg/L [36]. Lignin is the most abundant of these suspended solids, accounting for >90% of the total suspended matter in pulp and paper wastewater [37].

Carbohydrates: Carbohydrates are another major pollutant in pulp and paper wastewater, which mainly include sugars, acetic acids, and carboxylic acids, and are primarily derived from cellulose, hemicellulose, and other substances decomposed during the paper process. Furthermore, some additives, such as bleaching and auxiliary agents, are used in the paper process, which leave a certain amount of carbohydrates in wastewater [38]. The carbohydrate content in pulp and paper wastewater is generally between 100 and 500 mg/L.

Ammonia nitrogen: Ammonia nitrogen is an important pollutant in pulp and paper wastewater, and its content lies generally between 50 and 200 mg/L. The content is mainly derived from added preservatives, pulp agents, lye, and other products. Pulp and paper wastewater contains ammonia and nitrogen compounds such as ammonia, amino acids, and amino sugars. The concentration of these nitrogen-containing substances can exceed 50 mg/L [39].

Cyanide: Cyanide is an important pollutant in pulp and paper wastewater. Its content typically ranges between 5 and 20 mg/L. It is primarily generated from the addition of bleaching agents and additives [40].

Sulfate compound: Sulfuric acid is a common reagent used to adjust the acidity during pulp and paper production because of its low price. Aluminum sulfate (namely, alum and bauxite) exhibits pH regulation, flocculation, media, and other effects and is also commonly used in pulp and paper production. The use of sulfuric acid and aluminum sulfate increases the concentration of sulfate in sewage [41]. Sulfate compound content in pulp and paper wastewater is generally between 100 and 500 mg/L.

Phosphate: Pulp and paper wastewater contains phosphate substances such as calcium and sodium phosphate. The content of these phosphor sources can reach >20 mg/L [10].

The chemical pollutants in pulp and paper wastewater are primarily suspended matter, carbohydrates, ammonia nitrogen, cyanide, sulfate, and phosphate compounds, among which the content of organic and suspended matter is high and can reach tens to hundreds of specific gravity [42]. In addition, other substances, such as metal ions and dyes, may impact water quality. Therefore, to effectively control chemical pollution by pulp and paper wastewater, effective control measures must be undertaken, such as reducing the use of raw materials, improving related technologies and wastewater treatment processes, and adopting effective purification technologies [43].

3.2. Microbial Pollution Characteristics

Pulp and paper wastewater is a type of pollutant that contains a large amount of organic and inorganic matter and microorganisms. Microbial pollution characteristics refer to the polluting biological substances present in pulp and paper wastewater as well as the type, quantity, and activity of these biological substances [40].

Bacteria, fungi, and viruses are the most common microorganisms found in pulp and paper wastewater, with bacteria being the most abundant, followed by fungi, and viruses being the least abundant. Bacteria in pulp and paper wastewater primarily include *Bacillus*, *Proteus*, anaerobic bacteria, *Escherichia aerogenes*, *Escherichia coli*, *Pseudomonas aeruginosa*, *Staphylococcus aureus*, *Klebsiella pneumoniae*, and *Vibrio cholerae* et al. [44]. These bacteria are highly active and can decompose organic matter to produce pollutants.

In addition to bacteria, pulp and paper wastewater contains several fungi, such as yeast. Yeast can generate a large amount of heat through reproduction, thereby affecting the temperature of the wastewater; it can also metabolize harmful toxins, causing pollution of the environment [45]. Further, the wastewater contains several other fungi, primarily including various molds and mold spores. These fungi can produce a large amount of organic acids, reduce the pH level of the wastewater, and produce various toxic substances, thereby polluting the environment [46].

Pulp and paper wastewater may contain many viruses, which can enter the human body through wastewater, causing infectious diseases and health hazards [47].

The microbial pollution characteristics of pulp and paper wastewater primarily refer to microorganisms such as bacteria, yeast, fungi, and viruses contained in it as well as their types, quantities, and activities. These microorganisms can produce several pollutants and cause infectious diseases, posing health hazards [48].

4. Common Treatment Methods for Pulp and Paper Wastewater

4.1. Physical Treatment

Physical treatment refers to the use of physical methods to remove pollutants from pulp and paper wastewater. Commonly used physical treatment methods for pulp and paper wastewater include sedimentation, filtration, centrifugation, flotation, and membrane filtration [49]. Sedimentation is a simple and economical method for removing suspended solids from wastewater. The principle of sedimentation is to use gravity to separate suspended matter from the wastewater. Adding flocculants or coagulants to the wastewater can improve settling processes [50]. In the flocculation precipitation method, a flocculant forms coagulation agents that are used to remove suspended particles, colloidal macromolecules, and small pollutants from water through various mechanisms. Currently, this method is gaining popularity for the tertiary treatment of pulp and paper wastewater. Kim et al. used polyaluminum chloride (PAC) as a coagulant and cationic polyacrylamide (PAM) as a flocculant over a pH range of 2–10 to treat paper wastewater through the coagulation–flocculation process. The optimal conditions for coagulation–flocculation were found to be as follows: PAC concentration of 3689 mg/L, c-PAM concentration of 39.9 mg/L, and pH of 5.4. This suggests that coagulation–flocculation may be an effective pretreatment process for paper wastewater prior to the biological treatment [51]. Filtration is the process of removing suspended solids from wastewater using filtering media. Commonly used filtering media are sand, gravel, and activated carbon. The addition of flocculants or coagulants to wastewater can improve the filtration process [52]. Centrifugation is the process of separating suspended solids from wastewater using centrifugal force. The flotation process uses air bubbles to separate suspended solids from wastewater. Further, the membrane filtration process uses membranes to separate suspended solids from wastewater. Physical treatment can effectively remove suspended solids and organic matter from wastewater; however, the removal effect on toxic substances is unsatisfactory. A specialized film is used to separate the solute from the solvent, enabling the solute in the solution to permeate through the film and accomplish the separation of the solvent. Tonni et al. conducted an investigation into the feasibility of membrane filtration processes, including reverse osmosis (RO), for treating white water generated during thermal mechanical pulp in pulp and paper mills. The study revealed that membrane treatment could effectively eliminate SiO_2 from wastewater samples. When subjected to initial concentrations of 200 mg/L and pressures of 11 and 10 bar, ultrafiltration or nanofiltration membranes could be used for 12 h to remove 80% and 85% of SiO_2, respectively. RO, under the same conditions, almost entirely eliminated targeted contaminants at 10 bar. The treated wastewater produced through membrane filtration complied with local legislation standards, and the wastewater contents are less than 50 mg [53,54]. The advantages of physical treatment methods are high treatment efficiency and good treatment effect; however, there are also certain disadvantages, such as the generation of a large amount of waste liquid during the treatment process and high treatment costs [55].

4.2. Chemical Treatment

Chemical treatment refers to the use of chemicals to remove pollutants from wastewater. Commonly used chemical treatment methods for pulp and paper wastewater include coagulation, precipitation, oxidation, and adsorption [56]. Coagulation involves using a coagulant to destabilize and agglomerate suspended solids in wastewater and then separating them from the wastewater. Commonly used coagulants include inorganic substances such as ferric chloride and aluminum sulfate and organic substances such as polyacrylamide [57]. Sedimentation is the process of settling suspended solids in wastewater using chemicals [58]. Commonly used precipitants are lime and soda ash. Oxidation involves the use of oxidants to oxidize organic matter in wastewater, and commonly used oxidants are chlorine and ozone. In advanced oxidation disposal method, advanced oxidation reactions are used to eliminate pollutants from industrial wastewater. Strong oxidants containing highly active hydroxyl radicals are introduced into the wastewater, causing advanced oxidation reactions between oxygen anion free radicals and organic pollutants in the wastewater. This process gradually transforms them into safe and harmless inorganic compounds, purifying the wastewater. The method has shown favorable results in treating industrial wastewater with heavy water pollution, particularly in the pulp and paper industry [40]. Adsorption involves the use of adsorbents to adsorb pollutants in wastewater, and commonly used adsorbents include activated carbon and zeolite [23,59]. Adsorption is achieved through a chemical reaction occurring between the surface of an adsorbent and contaminants in the wastewater. In the chemical adsorption process, the adsorbent's surface usually features unsaturated bonds or active functional groups, allowing for chemical reactions with wastewater pollutants, forming chemical bonds or complexes. The principle underlying wastewater treatment via adsorption is based on the high affinity of adsorbents for pollutants in the wastewater. The choice of adsorbents is closely related to the characteristics, concentration, and desired discharge standards of the pollutants in wastewater. Various adsorbents exhibit varying degrees of selectivity for different pollutants. Adsorbents usually possess a large specific surface area and pore structure, providing numerous adsorption sites, which, in turn, increase adsorption capacity and efficiency. The advantages of the adsorption method for wastewater treatment include simplicity, effectiveness against various pollutants, and ability to regenerate adsorbents. However, there are certain limitations in treating wastewater by adsorption, such as the selection of appropriate adsorbents and the challenges of regeneration. Therefore, practical applications require careful consideration of wastewater characteristics and requirements, selection of suitable adsorbents, and optimization of treatment conditions to achieve optimal results. The advantages of chemical treatment include high treatment efficiency and good treatment effect; however, there are some disadvantages, such as poor removal effect of heavy metal ions, generation of a large amount of waste liquid during the treatment process, and high treatment costs.

4.3. *Biological Treatment*

Biological treatment is classified into aerobic sludge and anaerobic biological methods based on the characteristics of the microorganisms used for the treatment.

4.3.1. Aerobic Sludge Method

The Aerobic sludge method is commonly used for treating pulp and paper wastewater. It primarily includes biofilm, biofilter, bioprecipitation, and activated sludge methods [60,61]. Further, it uses the degradation mechanism of aerobic microorganisms to degrade organic matter in wastewater into carbon dioxide, water, and inorganic salts, thereby achieving the purpose of purifying wastewater [62]. The activated sludge and biofilm methods also use microorganisms to biodegrade organic matter in wastewater [63,64]. The advantages of biological treatment include low investment costs, low operating costs, and effective treatment. However, there are some disadvantages, such as lengthy treatment processes and difficulty in degrading heavy metals and antibiotics in wastewater [65].

4.3.2. Anaerobic Biological Method

The anaerobic biological method is suitable for treating pulp and paper wastewater with medium to high levels of pollutant concentrations. A wastewater treatment reaction device—upflow anaerobic sludge blanket (UASB reactor)—is employed, which is widely used in wastewater treatment in the pulp and paper industry around the world. Wastewater is introduced into the UASB reactor's bottom as evenly as possible. As the wastewater ascends through a sludge bed comprising granular or flocculent sludge, anaerobic reactions occur upon contact with sludge particles. The biogases produced in the anaerobic state (mainly methane and carbon dioxide) promote internal circulation, which is beneficial for the formation and maintenance of granular sludge. Some of the gases produced in the sludge layer adhere to the sludge particles. These gases, both attached and unattached, rise to the top of the reactor. When the sludge rises to the surface, it encounters the bottom of the gas emitter in the three-phase reactor. This interaction leads to the degassing of the sludge floc attached to the bubbles. Once the bubbles are released, the sludge particles settle back onto the sludge bed. Meanwhile, the attached and unattached gases are collected in the gas collecting chamber located within the three-phase separator situated at the top of the reactor. In current engineering practice, the methods of choosing an upflow anaerobic filter and anaerobic biofilter anaerobic baffle reactor are the most frequently employed [66].

4.4. Advanced Integrated Processing Technology

In addition to the above technologies, other technologies for the treatment and application of pulp and paper wastewater, such as electrochemical and ultrasonic treatment, have emerged in recent years [67]. Electrochemical and sonication treatments are the processes of removing pollutants from wastewater using electricity and ultrasonic waves, respectively [68]; these methods are gradually being employed in the industry.

Comprehensive treatment is the most commonly used method in the treatment of pulp and paper wastewater [69]. It primarily uses biological, chemical, and physical treatments as well as other technologies to separate harmful substances such as organic matter and heavy metals from wastewater [70]. The ozone biological method involves directing water, after undergoing ozone treatment, into a biological filter from the bottom, where aeration is provided as needed. This biological filter is equipped with an activated carbon filler as a biological carrier, allowing for the further removal of residual COD through biological contact oxidation. Cui et al. assessed the suitability of the ozone biological method for treating wastewater with low COD concentrations, such as raw water with a COD of only 50 mg/L. This method can achieve a COD level of <30 mg/L. In cases where raw water has a high COD concentration and strict discharge standards apply, a two-stage ozone biological series treatment can be performed. This approach eliminates the need for additional chemicals, ensures safety, and minimizes labor requirements. Melchiors et al. evaluated the efficiency of a physicochemical pretreatment combination, combining the coagulation–flocculation–sedimentation (CFS) process with the Fenton advanced oxidation process for treating wastewater generated from industrial chemithermal–mechanical pulp (CTMP) in Brazil. When the CFS and Fenton processes were combined for CTMP wastewater treatment, the total organic carbon removal rate reached 95%, the COD removal rate reached 61%, and the lignin removal rate reached 76% [40]. The advantages of comprehensive treatment include good treatment effects and high treatment efficiency; however, there are some disadvantages, such as high treatment costs [71]. At present, the commonly used pulp and paper wastewater treatment methods and their performance comparison are shown in Table 3.

Table 3. Common methods used for treating pulp and paper wastewater.

Handling Method	Performance Comparison
Flocculation precipitation method	The method is simple, the investment is not high, the operating cost is low, but it is difficult to reach the standard [51].
Advanced oxidation treatment	The method is simple, there is no sludge, the safety is poor, the investment is not high, and the operation cost is high [40].
Adsorption treatment	The investment is reasonable, but the regeneration is troublesome and the operation cost is high [59].
Aerobic sludge process	Less investment, low operating cost, good treatment effect. However, the treatment process is long, and it is difficult to degrade heavy metals and antibiotics in wastewater [61].
Biological ozone method	High investment, simple operation, reasonable operating cost, more suitable for upgrading [70].
A combination of coagulation–flocculation–precipitation process and Fenton advanced oxidation process	The operation is stable, the operation cost is low, but the operation amount is large, the safety is poor, and the sludge amount is large [45].

5. Utilization of Pulp and Paper Wastewater

At present, wastewater generated during the pulp and paper process is typically used as a resource after treatment. Common utilization technologies for pulp and paper wastewater include recycling and energy recovery [72]. Recycling refers to the process of using wastewater to produce paper or other products. Energy recovery is the process of using wastewater to generate electricity or heat.

5.1. Water Recycling

Pulp and paper wastewater has agricultural applications; treated pulp and paper wastewater can be used for agricultural irrigation. Organic matter, as well as nitrogen, phosphorus, and potassium in pulp and paper wastewater can provide nutrients to the plants, improve soil fertility, and promote crop growth [73]. Pulp and paper wastewater also has industrial applications; treated pulp and paper wastewater can be used for industrial purposes, which can reduce freshwater consumption as well as replace freshwater usage [74]. Wastewater from a pulp and paper workshop at Shandong Century Sunshine Paper Group was comprehensively treated. The process involved flocculation, primary anaerobic treatment, secondary aerobic treatment, coagulation, and precipitation. Some of the treated wastewater was further treated by sand filtration and subsequently used for the greening of the paper workshop and factory area. Excess wastewater that remains after treatment is discharged [75]. Treated pulp and paper wastewater can be used to power sewage treatment plants, wash equipment, clean streets, and other purposes [76].

5.2. Energy Recovery

Biogas produced in the anaerobic reaction treatment of pulp and paper wastewater is a good bioenergy resource. The biogas produced is used for power generation or as a substitute for coal and other energy sources, ensuring self-sufficiency in the wastewater treatment system. Any excess electrical energy is transmitted for use in the pulp and paper process. This approach achieves energy saving and reduction in energy consumption during wastewater treatment, ultimately leading to decreases carbon emissions. During the anaerobic reaction process, ~75% of COD is converted into methane gas. Gas production at 35 °C is calculated to be 0.4 m^3 of biogas per 1 kg of COD removed. The biogas output of an anaerobic reactor is at least 6500 m^3 per day [77]. Biogas produced during the anaerobic treatment process is collected and discharged via torch combustion, which pollutes the environment and wastes energy. Hence, the biogas is purified, and methane gas purified through biological desulfurization is used for hot air drying of paper machines, power generation, and factory heating [78]. Shandong Century Sunshine Paper Group employs this method for black liquor treatment. Ali et al. showed that using paper mill biosludge to produce biochar and bioenergy leads to substantial reduction of carbon emissions. The advantages of using paper mill biosludge as biochar, extend beyond mitigating climate change. The benefits include enhanced nutrient use efficiency and soil fertility, particularly in forests with poor soil quality, as well as improved soil acidification and soil health through heavy metal management [47].

5.3. Alkali Recovery

Many pulp and paper manufacturers perform alkali recovery because pulp and paper wastewater contains a high concentration of alkali. Alkali recovery technologies typically include combustion causticization, electrodialysis, and hydrothermal carbonization [79]. Among them, the combustion causticization method involves four production processes: extraction, evaporation, combustion, and causticization calcium oxide recovery. The combustion causticization method can only recover the alkali used in pulp, and a large amount of lignin is burned after black liquor is evaporated to dryness. The method involves high energy consumption and requires multiple pieces of evaporation equipment, making small- and medium-sized pulp mills unsuitable for the implementation of this method [80]. The electrodialysis method involves passing ions in black liquor from a low-concentration region to reach a high-concentration region through a semipermeable membrane. Under the action of a DC electric field, alkali can be recovered on the anode side. Further, the pH value of the cathode side is reduced due to the enrichment of hydrogen ions, and some lignin is precipitated; however, the recovery rate of this method is only 50% [81]. The hydrothermal carbonization method uses carbon dioxide as the acidifying agent, and organic impurities in black liquor are precipitated as carbonization raw materials in a closed environment. Compared with the traditional alkali recovery method, this method has low energy consumption, no wastewater and residue discharge, and can produce high-value-added activated carbon products [82]. Ji et al. demonstrated the feasibility of the direct causticization technology using reed black liquor with limestone as the bedding material (Figure 2). NaOH was successfully recovered through direct causticization. Approximately 91.2% of the sodium introduced into the system could be effectively recovered, with 87.4% of it converted into NaOH [83].

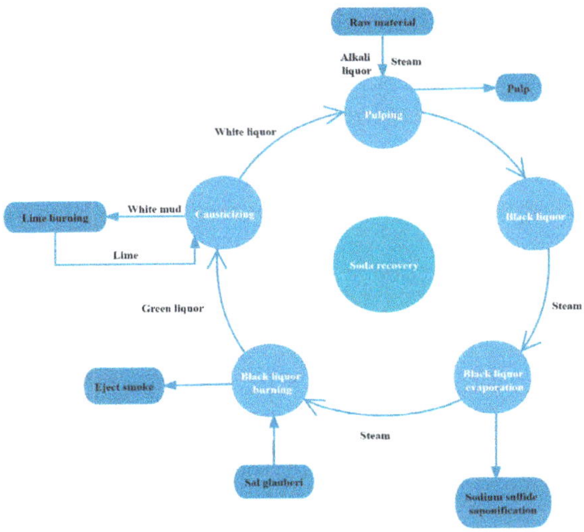

Figure 2. Flow chart of alkali recovery process in black liquor.

5.4. Lignin Recovery

The organic components in black liquor produced during the pulp process mainly include lignin and polysaccharides. During the cooking process, due to the action of sodium hydroxide, the ether bonds in lignin break and combine with hydroxide ions to form alkali lignin [12]. Alkali lignin is a natural polymer surfactant. After chemical modification, it can be used as a nontoxic renewable resource to replace petrochemical products; it can also be used as an adsorbent for heavy metal treatment, which has high application value. Alkali lignin can be precipitated by lowering the pH value. Common methods include adding acid to black liquor or introducing carbon dioxide. Jeroen et al. developed a pilot-scale procedure for mild alkaline pulp, followed by the recovery of lignin from black liquor. In a pilot-scale experiment, lignin recovery was achieved through acidification, enzyme treatment, and subsequent filtration. The study revealed that the purity of lignin was higher when a flocculant was employed [84]. The optimal condition for acid precipitation and dealkalization of lignin include a pH of <3, and the precipitation rate of up to 90% or higher can be achieved at a temperature of 25 °C. Figure 3 shows the lignin recovery process of the acid analysis of black liquor [85].

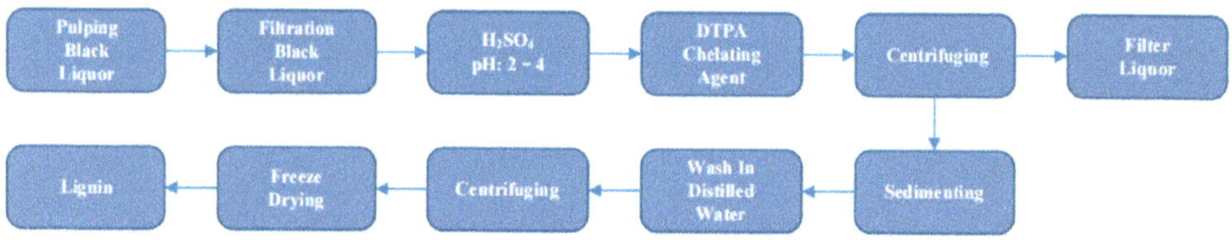

Figure 3. Lignin recovery process of the acid analysis of black liquor.

6. Sustainable Utilization of Pulp and Paper Wastewater

Sustainable utilization refers to minimizing environmental pollution, protecting natural resources, and realizing sustainable resource use in the case of limited resources and energy [86,87]. In the paper process, a large amount of wastewater is produced, which contains wood pulp, chemical additives, pigments, and other harmful substances. Effectively using this wastewater to reduce environmental pollution is an important concern in paper manufacturing [88].

First, technological transformation must be sought. Investment in technological transformation should be increased, and new technologies should be promoted and adopted, such as activated sludge treatment, wet dust removal, and ion exchange technology, to effectively reduce pollutant emissions [89]. It is also necessary to improve technical guidance for enterprises, raise environmental protection awareness among enterprises, encourage enterprises to improve technology, and effectively control wastewater discharge [89,90].

Further, the funding guarantee mechanism must be improved. Financial support for technological transformation should be increased and provided for enterprise transformation; the cost of enterprise transformation should also be reduced [91]. Furthermore, tax incentives should be offered to enterprises to encourage them to invest in environmental protection technologies to promote the development of wastewater treatment technologies [40].

Finally, an efficient regulatory system must be established. For different types of paper wastewater, corresponding discharge standards should be established. Further, an effective monitoring and testing system should be established, and strict penalties should be imposed on enterprises with excessive discharge [92]. In addition, an effective information-sharing mechanism should be established to promptly release information about pollution sources, and regular inspections of pollution sources should be conducted to ensure that enterprises comply with emission standards [93].

To develop an effective and sustainable use policy of pulp and paper wastewater, we must strengthen technological transformation, improve the fund guarantee mechanism, and establish an efficient supervision system [88,91]. Hence, by these discussed solutions, we can effectively control the discharge of pulp and paper wastewater, protect the environment, and effectively use resources [94].

7. Conclusions and Future Perspectives

With the development of society, the output of paper products—the products of the pulp and paper industry—is increasing, and paper products are inextricably linked to human life. Wastewater from the pulp and paper process must be safely treated and discharged. Various methods, such as advanced oxidation, membrane filtration, and activated sludge method, can be employed for paper wastewater treatment. Each of these methods is suited for different water quality conditions, and all these methods hold valuable research potential. In recent years, domestic and international researchers have increasingly focused on the combination of multiple processes for paper wastewater treatment. Thus, there is a need to explore the combinations of chemical, physical, and biological methods to maximize wastewater treatment efficacy and achieve cost efficiency.

In the 21st century, people's awareness of environmental protection is increasing, resulting in a paradigm shift; for example, using biological enzymes in the production process and all-biological pulp using high-efficiency biological enzymes. Recycling pulp and paper wastewater has become the future trend toward the green development of the pulp and paper industry. For example, manufacturers modify the pulp and paper process to recycle wastewater discharged during the process. If potassium hydroxide is used instead of sodium hydroxide in the pulp and paper process (there is no difference between the two in terms of pulp performance), discharged pulp and paper wastewater will contain sufficient potassium. Potassium fertilizer can be obtained from the discharged pulp and paper wastewater through secondary production, which not only offers economic benefits to factories but also reduces wastewater pollution. People created the first paper without the use of chemicals over 2000 years ago. With the advancement of modern biotechnology, highly efficient enzymes derived from organisms suitable for the pulp and paper industry have emerged. Because of their mild action conditions and high action efficiency, biological enzymes can effectively reduce energy consumption of the pulp and paper process and reduce wastewater pollution. We anticipate that by implementing new technologies and processes, the pulp and paper process will become more environmentally friendly and efficient in the future.

Author Contributions: K.L. and X.L. conceived the study. K.L., X.L., L.Y., Y.X. and R.W. wrote the draft of the manuscript. P.L. and J.W. critically reviewed the full manuscript content. All authors have read and agreed to the published version of the manuscript.

Funding: This work was supported by the National Key Research and Development Program of China (No. 2019YFC1905902); the Foundation (No. 202008) of Qilu University of Technology of Cultivating Subject for Biology and Biochemistry; the Foundation (No. 2022GH026) of International Technology Cooperation Project from Qilu University of Technology (Shandong Academy of Sciences); the Foundation (No. FWL2021065) of Shandong Provincial Key Laboratory of Biophysics; and the Foundation (No. 21YF5FA129) of Gansu Provincial Key R&D Program-Social Development. The funders had no role in study design, data collection and analysis, decision to publish, or preparation of the manuscript.

Data Availability Statement: The datasets supporting the results of this article are included within the article.

Conflicts of Interest: The authors declare no conflict of interest.

References

1. Wei, S.; Liu, K.; Ji, X.; Wang, T.; Wang, R. Application of Enzyme Technology in Biopulping and Biobleaching. *Cellulose* **2021**, *28*, 10099–10116. [CrossRef]
2. Guo, C. *Overview of the World Paper Industry in 2021*; China Pulp and Paper Magazines Publisher: Beijing, China, 2023; pp. 32–35.
3. Nations, Food and Agriculture Organization of the United. *Pulp and Paper Capacities*; Nations, Food and Agriculture Organization of the United: Roma, Italy, 2023.
4. China Paper Association. *China Pulp and Paper Industry: Annual Report 2022*; China Paper Association: Beijing, China, 2023; Volume 44, pp. 21–30+26.
5. Singh, A.K.; Chandra, R. Pollutants Released from the Pulp Paper Industry: Aquatic Toxicity and Their Health Hazards. *Aquat. Toxicol.* **2019**, *211*, 202–216. [CrossRef]
6. Singh, A.K.; Kumar, A.; Chandra, R. Environmental Pollutants of Paper Industry Wastewater and Their Toxic Effects on Human Health and Ecosystem. *Bioresour. Technol. Rep.* **2022**, *20*, 101250. [CrossRef]
7. Scholes, E.; Brook Carter, P.T.; Verheyen, T.V. Colloidal Carbon Interference in the Treatability of Pulp and Paper Wastewater by MBR. *J. Environ. Chem. Eng.* **2019**, *7*, 102943. [CrossRef]
8. Singh, G.; Arya, S.K. Utility of Laccase in Pulp and Paper Industry: A Progressive Step Towards the Green Technology. *Int. J. Biol. Macromol.* **2019**, *134*, 1070–1084. [CrossRef] [PubMed]
9. Yang, G.; Cheng, D.; Liu, Y.; Dongdong, C.; Zhang, M.; Zhao, H.; Sun, Y.; Huang, S.; Sao, J.; Xu, T.; et al. A Simple Method for Preparing Chelated Calcium Based on the Black Liquor Generated by the Pulp and Paper Industry. *Environ. Pollut. Bioavailab.* **2019**, *31*, 56–59. [CrossRef]
10. Jiao, G.J.; Ma, J.; Li, Y.; Jin, D.; Ali, Z.; Zhou, J.; Sun, R. Recent Advances and Challenges on Removal and Recycling of Phosphate from Wastewater Using Biomass-Derived Adsorbents. *Chemosphere* **2021**, *278*, 130377. [CrossRef] [PubMed]
11. Ramos, M.D.N.; Rangel, A.S.; Azevedo, K.S.; Melo, M.G.B.; Oliveira, M.C.; Watanabe, C.M.U.; Pereira, F.F.; Silva, C.M.; Aguiar, A. Characteristics and Treatment of Brazilian Pulp and Paper Mill Effluents: A Review. *Environ. Monit. Assess.* **2022**, *194*, 651. [CrossRef] [PubMed]

12. Li, N.; An, X.; Xiao, X.; An, W.; Zhang, Q. Recent Advances in the Treatment of Lignin in Papermaking Wastewater. *World J. Microbiol. Biotechnol.* **2022**, *38*, 116. [CrossRef]
13. Hay, J.X.W.; Wu, T.Y.; Ng, B.J.; Juan, J.C.; Jahim, J.M. Reusing Pulp and Paper Mill Effluent as a Bioresource to Produce Biohydrogen through Ultrasonicated Rhodobacter Sphaeroides. *Energy Convers. Manag.* **2016**, *113*, 273–280. [CrossRef]
14. Lin, Y.; Lin, H.; Chen, K.; Cai, T.; Liu, Z.; Wang, Y.; Chhuond, K. Research Progress on Wastewater Treatment Methods and Technologies in Paper Industry. *IOP Conf. Ser. Earth Environ. Sci.* **2019**, *358*, 022042. [CrossRef]
15. Mandeep; Kumar Gupta, G.; Shukla, P. Insights into the Resources Generation from Pulp and Paper Industry Wastes: Challenges, Perspectives and Innovations. *Bioresour. Technol.* **2020**, *297*, 122496. [CrossRef]
16. Li, Y.; Zhu, T.; Yang, H.; Nong, G. Characteristics of Calcium Lignin from Pulping Waste Liquor and Application for the Treatment Middle-Stage Wastewater of Paper Making. *Environ. Technol.* **2023**, *44*, 695–707. [CrossRef] [PubMed]
17. Castro, A.J.G.; Baptista, I.E.; de Moura, K.R.S.; Padilha, F.; Tonietto, J.; de Souza, A.Z.P.; Soares, C.H.L.; Silva, F.; Van Der Kraak, G. Exposure to a Brazilian Pulp Mill Effluent Impacts the Testis and Liver in the Zebrafish. *Comp. Biochem. Physiol. C Toxicol. Pharmacol.* **2018**, *206–207*, 41–47. [CrossRef] [PubMed]
18. Camcioglu, S.; Ozyurt, B.; Hapoglu, H. Effect of Process Control on Optimization of Pulp and Paper Mill Wastewater Treatment by Electrocoagulation. *Process Saf. Environ. Prot.* **2017**, *111*, 300–319. [CrossRef]
19. Kumar Singh, S.; Sharma, C.; Maiti, A. Forward Osmosis to Treat Effluent of Pulp and Paper Industry Using Urea Draw-Solute: Energy Consumption, Water Flux, and Solute Flux. *Sep. Purif. Technol.* **2021**, *278*, 119617. [CrossRef]
20. Mandeep; Liu, H.; Luo, J.; Shukla, P. Effluents Detoxification from Pulp and Paper Industry Using Microbial Engineering and Advanced Oxidation Techniques. *J. Hazard. Mater.* **2020**, *398*, 122998. [CrossRef]
21. Jahan, M.S.; Haris, F.; Rahman, M.M.; Samaddar, P.R.; Sutradhar, S. Potassium Hydroxide Pulping of Rice Straw in Biorefinery Initiatives. *Bioresour. Technol.* **2016**, *219*, 445–450. [CrossRef]
22. Romo, J.; Chaudhary, M.; Walker, T.R. Baseline Assessment of Contaminants in Marine Biota Prior to Remediation of Industrial Effluent Impacted Sediments in a Former Tidal Estuary in Nova Scotia, Canada. *Mar. Pollut. Bull.* **2019**, *145*, 641–648. [CrossRef]
23. Kumar, A.; Singh, A.K.; Chandra, R. Comparative Analysis of Residual Organic Pollutants from Bleached and Unbleached Paper Mill Wastewater and Their Toxicity on *Phaseolus aureus* and *Tubifex tubifex*. *Urban Water J.* **2020**, *17*, 860–870. [CrossRef]
24. Andersson, E.; Thollander, P. Key Performance Indicators for Energy Management in the Swedish Pulp and Paper Industry. *Energy Strategy Rev.* **2019**, *24*, 229–235. [CrossRef]
25. Sachan, P.; Madan, S.; Hussain, A. Isolation and Screening of Phenol-Degrading Bacteria from Pulp and Paper Mill Effluent. *Appl. Water Sci.* **2019**, *9*, 100. [CrossRef]
26. Fan, J.; Cao, Y.; Li, T.; Li, J.; Qian, X.; Shen, J. Unmodified Starch Granules for Strengthening Mineral-Filled Cellulosic Fiber Networks Promoted by Starch Pretreatment with a Cationic Polymer Flocculant in Combination with the Use of an Anionic Microparticulate Material. *ACS Sustain. Chem. Eng.* **2015**, *3*, 1866–1872. [CrossRef]
27. Kumar, V.; Chopra, A.K. Fertigation With Agro-residue-Based Paper Mill Effluent on a High-Yield Spinach Variety. *Int. J. Veg. Sci.* **2015**, *21*, 69–97. [CrossRef]
28. Magnusson, B.; Ekstrand, E.M.; Karlsson, A.; Ejlertsson, J. Combining High-Rate Aerobic Wastewater Treatment with Anaerobic Digestion of Waste Activated Sludge at a Pulp and Paper Mill. *Water Sci. Technol.* **2018**, *77*, 2068–2076. [CrossRef] [PubMed]
29. Ospina-Betancourth, C.; Acharya, K.; Allen, B.; Head, I.M.; Sanabria, J.; Curtis, T.P. Valorization of Pulp and Paper Industry Wastewater Using Sludge Enriched with Nitrogen-Fixing Bacteria. *Water Environ. Res.* **2021**, *93*, 1734–1747. [CrossRef] [PubMed]
30. Costa, S.; Dedola, D.; Pellizzari, S.; Blo, R.; Rugiero, I.; Pedrini, P.; Tamburini, E. Lignin Biodegradation in Pulp-and-Paper Mill Wastewater by Selected White Rot Fungi. *Water* **2017**, *9*, 935. [CrossRef]
31. Li, T.; Tong, Z.; Meng, S.; Li, Y.C.; Gao, B.; Bayabil, H.K. Characterization of Residues from Non-Woody pulp Process and Its Function as Fertilizer. *Chemosphere* **2021**, *262*, 127906. [CrossRef]
32. Vashi, H.; Iorhemen, O.T.; Tay, J.H. Degradation of Industrial Tannin and Lignin from Pulp Mill Effluent by Aerobic Granular Sludge Technology. *J. Water Process Eng.* **2018**, *26*, 38–45. [CrossRef]
33. Ram, C.; Rani, P.; Gebru, K.A.; Mariam Abrha, M.G. Pulp and Paper Industry Wastewater Treatment: Use of Microbes and Their Enzymes. *Phys. Sci. Rev.* **2020**, *5*, 20190050. [CrossRef]
34. John, D.; Yesodharan, S.; Achari, V.S. Integration of Coagulation-Flocculation and Heterogeneous Photocatalysis for the Treatment of Pulp and Paper Mill Effluent. *Environ. Technol.* **2022**, *43*, 443–459. [CrossRef]
35. Rudakova, V.; Bolotova, K.; Rudakov, I.; Subbotina, E.; Chukhchin, D. Biomonitoring in the Control System of Biological Wastewater Treatment in Pulp and Paper Industry. *IOP Conf. Ser. Earth Environ. Sci.* **2019**, *263*, 012017. [CrossRef]
36. Yiin, C.L.; Ho, S.; Yusup, S.; Quitain, A.T.; Chan, Y.H.; Loy, A.C.M.; Gwee, Y.L. Recovery of Cellulose Fibers from Oil Palm Empty Fruit Bunch for Pulp and Paper Using Green Delignification Approach. *Bioresour. Technol.* **2019**, *290*, 121797. [CrossRef]
37. Ram, M.; Mondal, M.K. Comparative Study of Native and Impregnated Coconut Husk with Pulp and Paper Industry Waste Water for Fuel Gas Production. *Energy* **2018**, *156*, 122–131. [CrossRef]
38. Tan, L.; Liu, Z.; Zhang, T.; Wang, Z.; Liu, T. Enhanced Enzymatic Digestibility of Poplar Wood by Quick Hydrothermal Treatment. *Bioresour. Technol.* **2020**, *302*, 122795. [CrossRef] [PubMed]
39. Liu, X.; Li, Y.; Meng, Y.; Lu, J.; Cheng, Y.; Tao, Y.; Wang, H. Pulp Black Liquor-Based Polymer Hydrogel as Water Retention Material and Slow-Release Fertilizer. *Ind. Crops Prod.* **2021**, *165*, 113445. [CrossRef]

40. Esmaeeli, A.; Sarrafzadeh, M.-H.; Zeighami, S.; Kalantar, M.; Bariki, S.G.; Fallahi, A.; Asgharnejad, H.; Ghaffari, S.-B. A Comprehensive Review on Pulp and Paper Industries Wastewater Treatment Advances. *Ind. Eng. Chem. Res.* **2023**, *62*, 8119–8145. [CrossRef]
41. Domínguez-Robles, J.; Palenzuela, M.d.V.; Sánchez, R.; Loaiza, J.M.; Espinosa, E.; Rosal, A.; Rodríguez, A. Coagulation–Flocculation as an Alternative Way to Reduce the Toxicity of the Black Liquor from the Paper Industry: Thermal Valorization of the Solid Biomass Recovered. *Waste Biomass Valorization* **2019**, *11*, 4731–4742. [CrossRef]
42. Wang, Y.; Liu, J.Z.; Chen, C.; Cheng, J. Slurryability and Combustion Characteristics of Coal-Coking Wastewater-Slurry. *Can. J. Chem. Eng.* **2019**, *97*, 1803–1808. [CrossRef]
43. Buzuku, S.; Farfan, J.; Harmaa, K.; Kraslawski, A.; Kässi, T. A Case Study of Complex Policy Design: The Systems Engineering Approach. *Complexity* **2019**, *2019*, 764368. [CrossRef]
44. De Oliveira, V.S.; Castro, A.J.G.; Cesconetto, P.A.; de Souza, A.Z.P.; Junior, J.J.B.; de Oliveira Nuner, A.P.; Soares, C.H.L.; Van Der Kraak, G.; Silva, F. Triterpene Betulin May Be Involved in the Acute Effects of Pulp and Paper Mill Effluent on Testis Physiology in Zebrafish. *Toxicol. In Vitro* **2021**, *73*, 105147. [CrossRef] [PubMed]
45. Grötzner, M.; Melchiors, E.; Schroeder, L.H.; dos Santos, A.R.; Moscon, K.G.; de Andrade, M.A.; Martinelli, S.H.S.; Xavier, C.R. Pulp and Paper Mill Effluent Treated by Combining Coagulation-Flocculation-Sedimentation and Fenton Processes. *Water Air Soil Pollut.* **2018**, *229*, 364. [CrossRef]
46. Popy, R.S.; Nayeem, J.; Arafat, K.M.Y.; Rahman, M.M.; Jahan, M.S. Mild Potassium Hydroxide pulp of Straw. *Curr. Res. Green Sustain. Chem.* **2020**, *3*, 100015. [CrossRef]
47. Mohammadi, A.; Sandberg, M.; Venkatesh, G.; Eskandari, S.; Dalgaard, T.; Joseph, S.; Granström, K. Environmental Analysis of Producing Biochar and Energy Recovery from Pulp and Paper Mill Biosludge. *J. Ind. Ecol.* **2019**, *23*, 1039–1051. [CrossRef]
48. Makela, M.; Forsberg, J.; Soderberg, C.; Larsson, S.H.; Dahl, O. Process Water Properties from Hydrothermal Carbonization of Chemical Sludge from a Pulp and Board Mill. *Bioresour. Technol.* **2018**, *263*, 654–659. [CrossRef]
49. Jaafarzadeh, N.; Omidinasab, M.; Ghanbari, F. Combined Electrocoagulation and Uv-Based Sulfate Radical Oxidation Processes for Treatment of Pulp and Paper Wastewater. *Process Saf. Environ. Prot.* **2016**, *102*, 462–472. [CrossRef]
50. Lal, K.; Garg, A. Effectiveness of Synthesized Aluminum and Iron Based Inorganic Polymer Coagulants for pulp Wastewater Treatment. *J. Environ. Chem. Eng.* **2019**, *7*, 103204. [CrossRef]
51. Kim, S.-C. Application of Response Surface Method as an Experimental Design to Optimize Coagulation–Flocculation Process for Pre-Treating Paper Wastewater. *J. Ind. Eng. Chem.* **2016**, *38*, 93–102. [CrossRef]
52. Aprianti, T.; Miskah, S.; Moeksin, R.; Sisnayati, S.; Nasir, S. Pb Removal in Pulp and Paper Industry Leachate Wastewater Using Activated Carbon-Ceramic Composite Adsorbent. *IOP Conf. Ser. Earth Environ. Sci.* **2019**, *298*, 012011. [CrossRef]
53. Kurniawan, T.A.; Othman, M.H.D.; Adam, M.R.; Goh, H.H.; Mohyudin, A.; Avtar, R.; Kusworo, T.D. Treatment of Whitewater from Pulp and Paper Industry Using Membrane Filtrations. *Chem. Pap.* **2022**, *76*, 5001–5010. [CrossRef]
54. Sun, S.; Liu, H.; Zhang, J.; Wang, W.; Xu, P.; Zhu, X.; Wang, Y.; Wan, S. Application of a Novel Coagulant in Reservoir Water Treatment in Qingdao. *Desalination Water Treat.* **2023**, *284*, 49–60. [CrossRef]
55. Jiang, Y.; Wu, Q.; Wei, Z.; Wang, J.; Fan, Z.; Pang, Z.; Zhu, Z.; Zheng, S.; Lin, X.; Chen, Y. Papermaking Potential of Pennisetum Hybridum Fiber after Fertilizing Treatment with Municipal Sewage Sludge. *J. Clean. Prod.* **2019**, *208*, 889–896. [CrossRef]
56. Bakraoui, M.; El Gnaoui, Y.; Lahboubi, N.; Karouach, F.; El Bari, H. Kinetic Study and Experimental Productions of Methane Production from Uasb Reactor Treating Wastewater from Recycled Pulp and Paper for the Continuous Test. *Biomass Bioenergy* **2020**, *139*, 105604. [CrossRef]
57. Lai, Y.H.; Sun, H.C.; Chang, M.H.; Li, C.C.; Shyu, J.G.; Perng, Y.S. Feasibility of Substituting Old Corrugated Carton Pulp with Thermal Alkali and Enzyme Pretreated Semichemical Mechanical Rice Straw Pulp. *Sci. Rep.* **2022**, *12*, 3493. [CrossRef]
58. Dai, Z.; Ma, Z.; Zhang, X.; Chen, J.; Ershadnia, R.; Luan, X.; Soltanian, M.R. An Integrated Experimental Design Framework for Optimizing Solute Transport Monitoring Locations in Heterogeneous Sedimentary Media. *J. Hydrol.* **2022**, *614*, 128541. [CrossRef]
59. Mahesh, S.; Garg, K.K.; Srivastava, V.C.; Mishra, I.M.; Prasad, B.; Mall, I.D. Continuous Electrocoagulation Treatment of Pulp and Paper Mill Wastewater: Operating Cost and Sludge Study. *RSC Adv.* **2016**, *6*, 16223–16233. [CrossRef]
60. Li, J.; Liu, Z.; Feng, C.; Liu, X.; Qin, F.; Liang, C.; Bian, H.; Qin, C.; Yao, S. Green, Efficient Extraction of Bamboo Hemicellulose Using Freeze-Thaw Assisted Alkali Treatment. *Bioresour. Technol.* **2021**, *333*, 125107. [CrossRef] [PubMed]
61. Dong, Y.; Yuan, H.; Ge, D.; Zhu, N. A Novel Conditioning Approach for Amelioration of Sludge Dewaterability Using Activated Carbon Strengthening Electrochemical Oxidation and Realized Mechanism. *Water Res.* **2022**, *220*, 118704. [CrossRef] [PubMed]
62. Wei, Z.; Li, Y.; Hou, Y. Quick Estimation for Pollution Load Contributions of Aromatic Organics in Wastewater from Pulp and Paper Industry. *Nord. Pulp Pap. Res. J.* **2018**, *33*, 568–572. [CrossRef]
63. Buchanan, I.D. Pulp and Paper Mill Effluent Management. *Water Environ. Res.* **2017**, *89*, 1417–1423. [CrossRef]
64. Hu, J.; Zhao, L.; Luo, J.; Gong, H.; Zhu, N. A Sustainable Reuse Strategy of Converting Waste Activated Sludge into Biochar for Contaminants Removal from Water: Modifications, Applications and Perspectives. *J. Hazard. Mater.* **2022**, *438*, 129437. [CrossRef] [PubMed]
65. Zhang, T.; Liu, S.; Li, H.; Ma, J.; Wang, X.; Shi, H.; Wang, Z.; Zhang, F.; Niu, M.; Guo, Y. One-Pot Preparation of Amphoteric Cellulose Polymers for Simultaneous Recovery of Ammonium and Dihydrogen Phosphate from Wastewater and Reutilizing as Slow-Release Fertilizer. *Eur. Polym. J.* **2022**, *171*, 111223. [CrossRef]

66. Kumar, A.; Srivastava, N.K.; Gera, P. Removal of Color from Pulp and Paper Mill Wastewater- Methods and Techniques—A Review. *J. Environ. Manag.* **2021**, *298*, 113527. [CrossRef] [PubMed]
67. Hooda, R.; Bhardwaj, N.K.; Singh, P. Brevibacillus Parabrevis Mtcc 12105: A Potential Bacterium for Pulp and Paper Effluent Degradation. *World J. Microbiol. Biotechnol.* **2018**, *34*, 31. [CrossRef]
68. Chu, H.; Wang, Z.; Liu, Y. Application of Modified Bentonite Granulated Electrodes for Advanced Treatment of Pulp and Paper Mill Wastewater in Three-Dimensional Electrode System. *J. Environ. Chem. Eng.* **2016**, *4*, 1810–1817. [CrossRef]
69. Institute, Shandong Environmental Protection Scientific Research and Design. *Technical Specification for Pulp and Paper Wastewater Treatment Engineering*; Institute, Shandong Environmental Protection Scientific Research and Design: Jinan, China, 2012.
70. Chen, J.; Fan, X.; Zhang, L.; Chen, X.; Sun, S.; Sun, R.C. Research Progress in Lignin-Based Slow/Controlled Release Fertilizer. *ChemSusChem* **2020**, *13*, 4356–4366. [CrossRef]
71. Feng, Z.; Chen, H.; Li, H.; Yuan, R.; Wang, F.; Chen, Z.; Zhou, B. Microwave-Assisted Koh Activated Lignite Semi-Coke for Treatment of Biologically Treated Wastewater from Pulp and Paper Mill. *J. Environ. Chem. Eng.* **2020**, *8*, 103924. [CrossRef]
72. Stoyanova, E.; Bochmann, G.; Couperus, A.; Fuchs, W. Enhanced Separation of the Organic Fraction from Paper Mill Effluent for Energy Recovery. *Waste Biomass Valorization* **2016**, *7*, 1031–1039. [CrossRef]
73. Dagar, S.; Singh, S.; Gupta, M. Integration of Pre-Treatment with UF/RO Membrane Process for Waste Water Recovery and Reuse in Agro-Based Pulp and Paper Industry. *Membranes* **2023**, *13*, 199. [CrossRef]
74. Saha, N.; Saba, A.; Saha, P.; McGaughy, K.; Franqui-Villanueva, D.; Orts, W.; Hart-Cooper, W.; Reza, M. Hydrothermal Carbonization of Various Paper Mill Sludges: An Observation of Solid Fuel Properties. *Energies* **2019**, *12*, 858. [CrossRef]
75. Coimbra, R.N.; Calisto, V.; Ferreira, C.I.A.; Esteves, V.I.; Otero, M. Removal of Pharmaceuticals from Municipal Wastewater by Adsorption onto Pyrolyzed Pulp Mill Sludge. *Arab. J. Chem.* **2019**, *12*, 3611–3620. [CrossRef]
76. Sudarshan, K.; Maruthaiya, K.; Kotteeswaran, P.; Murugan, A. Reuse the Pulp and Paper Industry Wastewater by Using Fashionable Technology. *Appl. Water Sci.* **2016**, *7*, 3317–3322. [CrossRef]
77. Lewis, R.; Cohen, J.; Awad, J.; Burger, H.; Marzouk, J.; Burch, G.; Lewis, D.M.; van Leeuwen, J.A. Study of the Impacts of Process Changes of a Pulp and Paper Mill on Aerated Stabilization Basin (Asb) Performance. *Chemosphere* **2018**, *211*, 767–774. [CrossRef]
78. Prabakar, D.; Suvetha, K.S.; Manimudi, V.T.; Mathimani, T.; Kumar, G.; Rene, E.R.; Pugazhendhi, A. Pretreatment Technologies for Industrial Effluents: Critical Review on Bioenergy Production and Environmental Concerns. *J. Environ. Manag.* **2018**, *218*, 165–180. [CrossRef]
79. Kaur, D.; Bhardwaj, N.K.; Lohchab, R.K. Environmental Aspect of Using Chlorine Dioxide to Improve Effluent and Pulp Quality During Wheat Straw Bleaching. *Waste Biomass Valorization* **2018**, *10*, 1231–1239. [CrossRef]
80. Ejraei, A.; Aroon, M.A.; Ziarati Saravani, A. Wastewater Treatment Using a Hybrid System Combining Adsorption, Photocatalytic Degradation and Membrane Filtration Processes. *J. Water Process Eng.* **2019**, *28*, 45–53. [CrossRef]
81. Sharma, K.; Pathak, M.; Kalita, S.; Bhattacharyya, K.G.; Sen Sarma, N.; Devi, A. Sequential Treatment of Paper Mill Effluent with Modified Fenton Oxidation and Bioflocculation. *Environ. Dev. Sustain.* **2019**, *22*, 5425–5442. [CrossRef]
82. Tong, X.; Shen, W.; Chen, X.; Corriou, J.-P. Qualitative and Quantitative Analysis of Gaseous Pollutants for Cleaner Production in Pulp and Paper Mills. *J. Clean. Prod.* **2018**, *198*, 1066–1075. [CrossRef]
83. Ji, X.; Bie, R.; Chen, P.; Gu, W. Reed Black Liquor Combustion in Fluidized Bed for Direct Causticization with Limestone as Bed Material. *Energy Fuels* **2016**, *30*, 5791–5798. [CrossRef]
84. Lauwaert, J.; Stals, I.; Lancefield, C.S.; Deschaumes, W.; Depuydt, D.; Vanlerberghe, B.; Devlamynck, T.; Bruijnincx, P.C.A.; Verberckmoes, A. Pilot Scale Recovery of Lignin from Black Liquor and Advanced Characterization of the Final Product. *Sep. Purif. Technol.* **2019**, *221*, 226–235. [CrossRef]
85. Liew, Y.X.; Chan, Y.J.; Manickam, S.; Chong, M.F.; Chong, S.; Tiong, T.J.; Lim, J.W.; Pan, G.T. Enzymatic Pretreatment to Enhance Anaerobic Bioconversion of High Strength Wastewater to Biogas: A Review. *Sci. Total Environ.* **2020**, *713*, 136373. [CrossRef]
86. Myllymäki, P.; Pesonen, J.; Nurmesniemi, E.-T.; Romar, H.; Tynjälä, P.; Hu, T.; Lassi, U. The Use of Industrial Waste Materials for the Simultaneous Removal of Ammonium Nitrogen and Phosphate from the Anaerobic Digestion Reject Water. *Waste Biomass Valorization* **2019**, *11*, 4013–4024. [CrossRef]
87. Trianni, A.; Negri, M.; Cagno, E. What Factors Affect the Selection of Industrial Wastewater Treatment Configuration? *J. Environ. Manag.* **2021**, *285*, 112099. [CrossRef] [PubMed]
88. Mosaddeghi, M.R.; Pajoum Shariati, F.; Vaziri Yazdi, S.A.; Nabi Bidhendi, G. Application of Response Surface Methodology (Rsm) for Optimizing Coagulation Process of Paper Recycling Wastewater Using Ocimum Basilicum. *Environ. Technol.* **2020**, *41*, 100–108. [CrossRef] [PubMed]
89. Khan, S.A.R.; Ponce, P.; Yu, Z.; Golpîra, H.; Mathew, M. Environmental Technology and Wastewater Treatment: Strategies to Achieve Environmental Sustainability. *Chemosphere* **2022**, *286*, 131532. [CrossRef]
90. Vashi, H.; Iorhemen, O.T.; Tay, J.H. Aerobic Granulation: A Recent Development on the Biological Treatment of Pulp and Paper Wastewater. *Environ. Technol. Innov.* **2018**, *9*, 265–274. [CrossRef]
91. Lyu, Y.; Ye, H.; Zhao, Z.; Tian, J.; Chen, L. Exploring the Cost of Wastewater Treatment in a Chemical Industrial Park: Model Development and Application. *Resour. Conserv. Recycl.* **2020**, *155*, 104663. [CrossRef]
92. Myllymäki, P.; Pesonen, J.; Romar, H.; Hu, T.; Tynjälä, P.; Lassi, U. The Use of Calcined Paper Mill Sludge as a Chemical Precipitant in the Simultaneous Removal of Ammonium and Phosphate-Paper Mill Waste Recycling and Reuse. *Desalination Water Treat.* **2020**, *194*, 459–467. [CrossRef]

93. Yuliani, G.; Chaffee, A.L.; Garnier, G. Uv-Induced Colour Generation of Pulp and Paper Mill Effluents as a Proxy of Ligno-Cellulosic Biorefinery Wastewater. *J. Water Process Eng.* **2019**, *29*, 100781. [CrossRef]
94. Jiang, J.-G.; Yuan, T.-Q.; Wang, S.-F.; Liu, S.-J.; Shi, X.-D.; Zheng, L.; Sun, R.-C. Upgrading Traditional Pulp Mill into Biorefinery Platform: Wheat Straw as a Feedstock. *ACS Sustain. Chem. Eng.* **2018**, *6*, 15284–15291. [CrossRef]

MDPI AG
Grosspeteranlage 5
4052 Basel
Switzerland
Tel.: +41 61 683 77 34

Water Editorial Office
E-mail: water@mdpi.com
www.mdpi.com/journal/water

www.ingramcontent.com/pod-product-compliance
Lightning Source LLC
Chambersburg PA
CBHW062047220926
51850CB00010B/2684
* 9 7 8 3 7 2 5 8 5 6 8 7 9 *